—Tasks and Rubrics for—
Balanced Mathematics Assessment

in Primary and Elementary Grades

Tasks and Rubrics for Balanced Mathematics Assessment

in Primary and Elementary Grades

Judah L. Schwartz

Joan M. Kenney

CORWIN PRESS
A SAGE Company
Thousand Oaks, CA 91320

Copyright © 2008 by Corwin Press

All rights reserved. When forms and sample documents are included, their use is authorized only by educators, local school sites, and/or noncommercial or nonprofit entities that have purchased the book. Except for that usage, no part of this book may be reproduced or utilized in any form or by any means, electronic or mechanical, including photocopying, recording, or by any information storage and retrieval system, without permission in writing from the publisher.

The Inch Worm from the Motion Picture HANS CHRISTIAN ANDERSEN. By Frank Loesser © 1951, 1952 (Renewed) FRANK MUSIC CORP. All Rights Reserved

For information:

Corwin Press
A SAGE Company
2455 Teller Road
Thousand Oaks, California 91320
www.corwinpress.com

SAGE Ltd.
1 Oliver's Yard
55 City Road
London, EC1Y 1SP
United Kingdom

SAGE India Pvt. Ltd.
B 1/I 1 Mohan Cooperative
Industrial Area
Mathura Road, New Delhi 110 044
India

SAGE Asia-Pacific Pte. Ltd.
33 Pekin Street #02-01
Far East Square
Singapore 048763

Library of Congress Cataloging-in-Publication Data

Schwartz, Judah L., 1934-
Tasks and rubrics for balanced mathematics assessment in primary and elementary grades / Judah L. Schwartz and Joan M. Kenney.
    p. cm.
Includes bibliographical references.
ISBN 978-1-4129-5730-4 (cloth)
ISBN 978-1-4129-5731-1 (pbk.)
    1. Mathematics--Study and teaching (Elementary)—United States—Evaluation. 2. Mathematical ability—Testing. I. Kenney, Joan M., 1937- II. Title.

QA135.6.S4363 2008
   372.7—dc22   2007040312

This book is printed on acid-free paper.

07 08 09 10 11 10 9 8 7 6 5 4 3 2 1

Acquisitions Editor: Cathy Hernandez
Editorial Assistants: Megan Bedell
Production Editor: Appingo Publishing Services
Cover Designer: Monique Hahn
Graphic Designer: Scott Van Atta

Contents

Preface   vii
Acknowledgments   ix
About the Authors   xi

1. **Introduction to Balanced Assessment**   1

2. **Number and Quantity**   15
    Primary:
        Add-Rings   18
        Birthday Cupcakes   23
        TV Shows   27
        A Very Long Hallway   30
        Dot-to-Dot   34
        Measuring the Marigolds   39
        Table Talk   43
        Leopard's Leap   50
        Add 'Em Up   53
    Elementary:
        Multiplication Rings   59
        Fermi Four   64
        Broken Calculators   67
        The Trouble With Tables   70
        Broken Measures   76
        Counting Off   80
        Network News   84
        Piece of String   89

3. **Shape and Space**   95
    Primary:
        Grassy Parks   98
        Stickers   102
        Shirts in the Mirror   105
    Elementary:
        Does It Fit?   112
        Mirror, Mirror   115
        Gardens of Delight   120

4. Pattern and Function — 123

Primary:
- Wall Design — 126
- Calendar Moves — 132

Elementary:
- Make a Map — 144
- Coding the Alphabet — 152

5. Chance and Data — 155

Primary:
- Beach Day — 157
- We Scream for Ice Cream — 160

Elementary:
- Mixed-Up Socks — 165
- Measure Me — 170

6. Arrangement — 175

Primary:
- Postal Puzzles — 178
- Shares — 182

Elementary:
- Valentine Hearts — 185
- Millie & Mel's — 189

References — 193
Index — 195

Preface

Starting in the early 1990s, with the support of the National Science Foundation (NSF), four groups of scholars, researchers, and teachers began to investigate alternative modes of assessment in mathematics under the common name of Balanced Assessment in Mathematics. These groups, at the University of California at Berkeley, Michigan State University, the Shell Mathematics Center at the University of Nottingham, and the Harvard Graduate School of Education, collaborated and competed over the next decade in evolving a variety of approaches to the problem of helping teachers and schools assess the effectiveness of their instruction in mathematics.

At Harvard, we worked on devising a way to formulate the content and methods of mathematics in a way that teachers and students would be comfortable with. We worked closely with teachers in classroom settings to be sure that the abstractions of the academic world translated into workable and, indeed, beneficial classroom products—products that had a coherence across the grades, across the various content domains of mathematics, and across the diverse interests of students and teachers.

The core of our formulation of the field of mathematics started from the recognition that each of the thousands of spoken languages on our planet structures language in the form of NOUN PHRASE–VERB PHRASE. Such universality implies something quite fundamental about the way we perceive and talk about the world. We seem to want to describe our environment and the events that occur in it in terms of OBJECTS (noun phrases) and ACTIONS (verb phrases). Does this observation about language shed any light on the way we might think about mathematics?

We can formulate the subject of mathematics in terms of mathematical objects (e.g., numbers, shapes, patterns) and mathematical actions (e.g., addition, multiplication, reflection, scaling) that are carried out on or by them. We have found that this formulation helps teachers put their own knowledge of the subject of mathematics into a perspective that allows them to communicate more effectively and more coherently with their students. Lest you fear that this formulation provides yet another thing to learn and understand, we show how the National Council of Teachers of Mathematic's (NCTM, 2006) *Curriculum Focal Points* fit directly and easily into this framework.

We have also paid particular attention to the complex nature of problem solving. Solving a problem is not a unitary task. Some problems make strong demands on a student's ability to formulate a problem mathematically, while other problems make strong technical demands on a student's ability to compute but little demand on the understanding necessary to know what to compute. Still other problems require the student to make inferences based on the results of their formulations and computations. Collapsing all of these dimensions of solving a problem into a single grade does little to help the teacher understand where a student's strengths and weaknesses might lie.

We hope that this book succeeds in conveying our excitement for balanced assessment to teachers and students of mathematics and makes clear the intertwined nature of instruction and assessment.

Acknowledgments

About half of the assessment tasks included in this Primary and Elementary collection were developed at Harvard University by the authors. The remainder were designed and written by other members of the Balanced Assessment Group at the Harvard Graduate School of Education, namely

Kevin A. Kelly

Teresa Sienkiewicz

Victor Steinbok

We wish to acknowledge their creativity and their ability to produce engaging assessments that provide comprehensive information about what students do and do not understand in their pursuit of mathematical learning.

This material is based upon work supported under National Science Foundation grants MDR-925902 and ESI-9736403 (subcontracts from the University of California at Berkeley and Michigan State University), and with supplemental funding from the American Academy of Arts and Sciences, the Boston Public Schools, the Cambridge Public Schools, and the Noyce Foundation.

We also would like to thank Bob Tinker and our other colleagues at the Concord Consortium for their support and encouragement.

PUBLISHER'S ACKNOWLEDGMENTS

Corwin Press gratefully acknowledges the contributions of the following individuals:

Kathryn McCormick
Math and Science Teacher
Gahanna Middle School East
Gahanna, OH

Marsha Sanders-Leigh
Program Director
The Center for Education Integrating Science, Mathematics,
 and Computing
Georgia Tech
Atlanta, GA

Melinda Webster
Fifth Grade Teacher
Lenoir City Elementary School
Lenoir, TN

About the Authors

 **Judah L. Schwartz** is currently Visiting Professor of Education and Research Professor of Physics and Astronomy at Tufts University, where he directs a large NSF-supported project on science education for middle school and elementary school teachers. He is also Emeritus Professor of Engineering Science and Education at the Massachusetts Institute of Technology and Emeritus Professor of Education at the Harvard Graduate School of Education. He was trained in theoretical physics and mathematics and did research for some years in the area of atomic physics. In the course of that research, he and his colleagues developed a variety of computer graphics techniques that proved to be useful in the teaching of mathematics and science. His current research interests include the design of microcomputer software environments to improve the teaching and learning of science and mathematics and the application of cognitive science techniques to the study of mathematics and science education. He has been a visiting professor at universities in France, Italy, and Israel; has consulted and lectured widely in this country and abroad; and has published extensively in the area of educational technology. He is the author or coauthor of many software environments including *The Semantic Calculator, The Algebraic Proposer, M-SS-NG L-NKS: A Game of Letters & Language, What Do You Do With A Broken Calculator?, The Geometric Supposer, Calculus Unlimited, Sir Isaac Newton's Games,* and *The Newtonian Sandbox.*

Judah has a long-standing interest in alternative modes of assessment and has edited reports titled "The Prices of Secrecy: The Social, Intellectual and Psychological Costs of Current Assessment Practice" and "Assessing Mathematics Understanding & Skills Effectively." Recent publications include a book-length case study of educational reform titled "The Geometric Supposer; What Is It a Case Of?" and "Software Goes to School: Teaching for Understanding in the Age of Technology." Judah may be contacted by e-mail at Judah.Schwartz@Tufts.edu.

Joan M. Kenney's professional career has encompassed a wide variety of experiences in the field of mathematics. She has worked as a research scientist, specializing in operations analysis and risk management; taught mathematics at the secondary and college level; and performed task modeling and pedagogical intervention in elementary and middle school classrooms. Joan served as the national evaluator for the NSF-sponsored *Assessment Community of Teachers* and *Connecting with Mathematics* projects, the *Instructional Leadership Academy* sponsored by the Council for Basic Education, and the *Digi-Block* program. She has delivered keynote addresses at several national and international conferences and has written extensively about mathematics education reform and assessment.

Joan recently retired from the Harvard Graduate School of Education, where for ten years she was the project coordinator and codirector of the Balanced Assessment in Mathematics Program. During that time she was involved in assessment task design, student performance evaluation, and outreach to community stake-holders. She also served on the Mathematics Task Force of the Massachusetts Board of Higher Education and on the original design committee for the Massachusetts Comprehensive Assessment System (MCAS). She continues to consult with school districts on issues of mathematics curriculum and classroom practice, and to provide professional development for teachers and administrators in the areas of mathematics content and assessment. Her book, *Literacy Strategies for Improving Mathematics Instruction*, was recently published by ASCD. Joan may be contacted by e-mail at Joan_Kenney@post.harvard.edu or kenneyjo@aol.com.

1

Introduction to Balanced Assessment

Assessing the mathematical performance of our students and the effectiveness of our mathematics instructional programs has become a major concern of the mathematics education community and the country as a whole. The National Council of Teachers of Mathematics (NCTM, 1995) has addressed that concern in its *Assessment Standards for School Mathematics*, which provides a set of six standards to guide the development of assessment instruments. This NCTM document makes clear, however, that it is a guide and not a "how-to" document. Guides are necessary but not sufficient, and teachers need different models of assessment that implement in concrete ways the principles set down in guidelines such as those offered by the NCTM.

The tasks in this collection were developed by the Balanced Assessment in Mathematics Program at the Harvard Graduate School of Education. The main goal of Balanced Assessment is to design assessments that can be used in classrooms throughout the nation—assessments that reflect the values of the mathematics reform movement as articulated in both NCTM's (1989) *Curriculum and Evaluation Standards for School Mathematics* and NCTM's (2006) *Curriculum Focal Points*. The assessments are intended to provide teachers, students, schools, and parents with useful information about how students and programs are doing with respect to those standards. The collection is balanced in regard to the mathematical content covered and the problem-solving skills utilized. The assessments have been created with a unique design philosophy that gets to the heart of the question "What is the structure of mathematics?"

WHAT IS MATHEMATICS ABOUT? THE STRUCTURE OF THE SUBJECT

As in many disciplines, it is possible to identify both content and process dimensions in the subject of mathematics. Unlike many subjects where most of the process dimension refers to general reasoning, problem-formulating, and problem-solving skills, the process dimension in mathematics refers to many skills that are mathematics-specific. As a result, many people tend to lump content and process together when speaking about mathematics, calling it all mathematics *content*.

It is important to maintain the distinction between content and process. We say this because we believe that this distinction reflects something very deep about the way humans approach mental activity of all sorts. All human languages have grammatical structures that distinguish between noun phrases and verb phrases. They use these structures to express the distinction between objects and the actions carried out by or on these objects.

The content-process distinction in mathematics is best described by the words *object* and *action*. What are the mathematical objects we wish to deal with? What are the mathematical actions that we carry out with these objects? We will try to answer these questions in a way that makes clear the continuity of the subject from the earliest grades through post-secondary mathematics. Since there are really very few discrete categories of mathematical objects and actions, we believe this approach offers a simple and clear way to view and organize the teaching and learning of mathematics, consistent with NCTM guidelines for both teaching and assessment standards.

THE OBJECTS OF MATHEMATICS

The first category of mathematical objects we consider is that of **number and quantity**. Indeed, elementary mathematics is largely about this group of objects and the actions we carry out with and on them. For this reason, a preponderance of the tasks in this collection are concerned with this topic. Some of the math objects included in number and quantity are

- integers (positive and negative whole numbers and zero);
- rationals (fractions, decimals, and all the integers);
- measures (length, area, volume, time, weight);
- real numbers (p, e, and all the rationals);
- complex numbers;
- vectors and matrices.

Along with number and quantity, we introduce very early a concern for another category of mathematical object, namely **shape and space**. Math objects investigated in this domain are

- topological spaces (concepts of connectedness and enclosure);
- metric spaces (with such shapes as lines/segments, polygons, circles, conic sections, etc.).

From the beginning we try to make students aware of **patterns** in the worlds of number and shape. In primary and elementary grades, patterns and sequences are closely aligned with arrangements. Pattern as a mathematical object matures into **function**, which is the central mathematical object of the subjects we call algebra and calculus. We include the following as math objects within the category **pattern and function**

- functions on real numbers (linear, quadratic, power, rational, periodic, transcendental);
- functions on shapes.

There are several other kinds of mathematical objects that have less prominent roles in the mathematics we expect our younger students to study. These include objects in the categories **chance and data** and **arrangement**.

The category **chance and data** is concerned with math objects such as

- relative frequency and probability;
- discrete and continuous data.

Some aspects of data collection, organization, and presentation can be done in the earliest grades, but little, if any, data analysis is performed. Notions of probability are not realistically addressable until late middle school.

In the earlier grades, **arrangement** tends to blend with the study of patterns of numbers and shapes. Some of the math objects to be considered in later elementary mathematics are

- permutations and combinations;
- graphs;
- networks, trees, and counting schemes.

THE ACTIONS OF MATHEMATICS

As previously mentioned, the process dimension of mathematics has many actions that are mathematics-specific. It also involves actions that are properly regarded as general problem-formulating, problem-solving, and reasoning skills—processes that are needed across all aspects of learning and living. We divide these skills into four categories:

- Modeling/Formulating
- Transforming/Manipulating
- Inferring/Drawing Conclusions
- Communicating

With the exception of communicating, each of these actions has aspects that are specific to mathematics and aspects that are not specific to mathematics, but that are quite general in nature. Some of these general and specific aspects are now listed.

<u>Modeling/Formulating</u>
   domain-general
      observation and evidence gathering
      necessary and/but not sufficient conditions
      analogy and contrast

  deciding, with awareness, what is important and what can be ignored

 domain-specific

  deciding, with awareness, what can be mathematized and then doing so

  formally expressing dependencies, relationships, and constraints

<u>Transforming/Manipulating</u>

 domain-general

  understanding "the rules of the game"

  understanding the nature of equivalence and identity

 domain-specific

  arithmetic computation

  symbolic manipulation in algebra and calculus

  formal proofs in geometry

<u>Inferring/Drawing Conclusions</u>

 domain-general

  shifting point of view

  testing conjectures

 domain-specific

  investigation of limiting cases

  investigation of symmetry and invariance

  investigation of "between-ness"

<u>Communicating</u>

  making a clear argument, both orally and in writing, using both prose and images

It is evident that there is no reasonable way to separate, nor should there be any interest in separating, the domain-specific and the domain-general aspects. We therefore come to the conclusion that it is better to parse the domain of mathematics as

object (number and quantity, shape and space, pattern and function, chance and data, arrangement)

and

action (including both domain-specific and domain-general actions)

rather than as

content (usually defined by "topics"—an undifferentiated mixture of objects and domain-specific actions)

and

process (i.e., domain-general actions), which is the usual procedure in mathematics education.

In transitioning to the objects × actions lens, it is helpful to begin by seeing the correspondence between the Balanced Assessment math objects and NCTM curriculum content. As shown in the following matrix, we can use the **objects × actions** structure to map the tasks and rubrics in this collection to NCTM's (2006, pp. 11–17) *Curriculum Focal Points*:

Figure 1.1

\multicolumn{2}{c}{**MAPPING OF BALANCED ASSESSMENT (BA) OBJECTS × ACTIONS to 2006 NCTM CURRICULUM FOCAL POINTS**}
CONTENT AREA/ MATH OBJECTS	LEARNING EXPECTATIONS
NCTM: Number and Operations, Measurement	NCTM: Develop an understanding of whole numbers, including concepts of correspondence, counting, cardinality, and comparison. Identify measurable attributes and compare objects by using these attributes. (Pre-K)
	Represent, compare, and order whole numbers, and join and separate sets. Order objects by measurable attributes. (K)
	Develop understandings of addition and subtraction and strategies for basic addition facts and related subtraction facts; also, whole number relationships, including grouping in tens and ones. (Gr. 1)
	Develop an understanding of the base-ten numeration system and place-value concepts; develop quick recall of addition facts and related subtraction facts and fluency with multi-digit addition and subtraction. Develop understanding of linear measurement and facility in measuring lengths. (Gr. 2)
	Develop understandings of multiplication and division and strategies for basic multiplication facts and related division facts; also fractions and fraction equivalence. (Gr. 3)
	Develop quick recall of multiplication facts and related division facts and fluency with whole number multiplication; also decimals, including the connections between fractions and decimals. Develop an understanding of area and determine the area of two-dimensional shapes. (Gr. 4)
	Develop an understanding of and fluency with division of whole numbers; also addition and subtraction of fractions and decimals. Describe three-dimensional shapes and analyze their properties, including volume and surface area. (Gr. 5)

Continued on next page.

Figure 1.1
Continued.

CONTENT AREA/ MATH OBJECTS	LEARNING EXPECTATIONS
BA: Number and Quantity	BA: Primary Grades: Students will grow in their capacity to • demonstrate a robust understanding of our numeration system; • demonstrate understanding of the conceptual meaning of addition and subtraction of whole numbers and integers, and the relationship between the two processes; • demonstrate a basic understanding of the meaning of simple fractional parts; • count anything in the world around us, and identify and measure continuous quantities such as length, area, and time. Elementary Grades: Students will grow in their capacity to • demonstrate computational facility with the four arithmetic operations on whole numbers and integers; • demonstrate understanding of the various meanings of multiplication and division of whole numbers and integers, and the relationship between the two processes; • make reasonable approximations of the results of arithmetic computations and estimates of measurement using standard and non-standard measures; • demonstrate understanding of the order properties of decimals and other rational fractions.
NCTM: Algebra	NCTM: Recognize and duplicate simple sequential patterns. (Pre-K) Identify, duplicate, and extend simple number patterns and sequential and growing patterns as preparation for creating rules that describe relationships. (K) Identify, describe, and apply number patterns and properties. (Gr. 1) Use number patterns to extend knowledge of properties of numbers and operations. (Gr. 2) Create and analyze patterns and relationships involving multiplication and division. (Gr. 3) Develop an understanding of the use of a rule to describe a sequence of numbers or objects. (Gr. 4) Use patterns, models, and relationships as contexts for writing and solving simple equations and inequalities; create graphs of simple equations; develop an understanding of the order of operations. (Gr. 5)

Figure 1.1
Continued.

CONTENT AREA/ MATH OBJECTS	LEARNING EXPECTATIONS
BA: Pattern and Function	BA: Primary Grades: Students will grow in their capacity to • recognize and extend numerical patterns or relationships; • recognize and extend spatial patterns; • enumerate and organize simple arrangements. Elementary Grades: Students will grow in their capacity to • express, both in words and symbolically, how one thing depends on another; • identify, generate, extend, and describe repetitive relationships, both numerical and spatial.
NCTM: Geometry	NCTM: Identify shapes and describe spatial relationships. (Pre-K) Identify, name, and describe a variety of shapes and spaces. (K) Compose and decompose geometric shapes. (Gr. 1) Estimate, measure, and compute lengths while solving problems involving data, space, and movement through space. (Gr. 2) Describe and analyze properties of two-dimensional shapes; introduce concepts of symmetry and congruence. (Gr. 3) Deepen and extend understanding of two-dimensional space by finding area; use transformations to design and analyze simple tilings and tessellations. (Gr. 4) Describe three-dimensional shapes and analyze their properties, including volume and surface area. (Gr. 5)
BA: Shape and Space	BA: Primary Grades: Students will grow in their capacity to • distinguish and name a variety of two-dimensional shapes; • demonstrate understanding of the symmetries of shapes. Elementary Grades: Students will grow in their capacity to • distinguish, name, and manipulate a variety of two- and three-dimensional shapes; • demonstrate understanding of the geometric concepts of distance, location, symmetry, similarity, translation, rotation, reflection, covering, projection, scaling, and tessellation.

Continued on next page.

Figure 1.1
Continued.

CONTENT AREA/ MATH OBJECTS	LEARNING EXPECTATIONS
NCTM: Data Analysis, Probability	NCTM: Use the attributes of objects (size, quantity, orientation, color, etc.) to describe, sort, and compare. (Pre–K) Sort objects and use one or more attributes to solve problems. (K) Solve problems involving measurements and data; represent discrete data in picture and bar graphs. (Gr. 1) Construct and analyze frequency tables, bar graphs, picture graphs, and line plots, and use them to solve problems. (Gr. 3) Apply understanding of place value to develop and use stem-and-leaf plots. (Gr. 4) Apply understanding of whole numbers, fractions, and decimals to construct and analyze double-bar and line graphs; use ordered pairs on a coordinate grid. (Gr. 5)
BA: Chance and Data, Arrangement	BA: <u>Primary Grades</u>: Students will grow in their capacity to • collect, organize, and display simple data sets; • make decisions based on provided data. <u>Elementary Grades</u>: Students will grow in their capacity to • organize and display discrete information in a variety of formats, including frequency tables, bar graphs, and line plots; • demonstrate competence in dealing with the effects of randomness and uncertainty in data collection; • enumerate and organize simple combinations and permutations.

Reprinted with permission from *Curriculum Focal Points for Prekindergarten through Grade 8 Mathematics*, copyright 2006 by the National Council of Teachers of Mathematics.

WEIGHTING OF TASKS

In order to approach the problem of designing balanced assessment packages in mathematics, one must have a clear view of the kind of understanding and the skills that we wish to assess in our students, and the ways in which the tasks we design elicit demonstrable evidence of these skills and understanding. In what follows, we shall describe how our view of the subject of mathematics—its objects and its actions—informs the design of tasks and the balancing of assessment packages.

Each task is classified according to domain, that is the mathematical objects that are prominent in the accomplishment of the task. Most of our tasks deal predominantly with a single sort of mathematical object, although some deal with two. Each task offers students an opportunity to demonstrate a variety of kinds of skill and understanding.

In order to score student performance on a task, one has to first analyze the task and decide on the nature of the demands that the task makes on the student. We considered the following four kinds of skill and understanding:

Modeling/Formulating: How well does the student take the presenting statement and formulate the mathematical problem to be solved? Some tasks make minimal demands along these lines. For example, a problem that asks students to calculate the length of the hypotenuse of a right triangle given the lengths of the two legs does not make serious demands along these lines. On the other hand, the problem of how many 3-inch diameter tennis balls can fit in a 3-inch × 4-inch × 10-inch rectangular or parallelepiped box, while exercising the same Pythagorean muscles in the solution, is rather different in the demands it makes on a student's ability to formulate problems.

Transforming/Manipulating: How well does the student manipulate the mathematical formalism in which the problem is expressed? This may mean adding two numbers or dividing one fraction by another, making a geometric construction, solving an equation or inequality, plotting graphs, or finding the derivative of a function. Most tasks will make some demands along these lines. Indeed, most traditional mathematics assessment consists of problems whose demands are primarily of this sort.

Inferring/Drawing Conclusions: How well does the student apply the results of his or her manipulation of the formalism to the problem situation that spawned the problem? Traditional assessments often pose problems that make little demand of this sort. For example, students may well be asked to demonstrate that they can multiply the polynomials $(x + 1)$ and $(x - 1)$, but may not be expected to notice (or understand) that the numbers one cell away from the main diagonal of a multiplication table always differ from perfect squares by exactly 1.

Similarly, a younger student might be asked to sort through what may seem to be contradictory pieces of information in order to develop a solution strategy.

Communicating: How well do students communicate to others what they have done in formulating the problem, manipulating the formalism, and drawing conclusions about the implications of their results?

Since we do not expect each task to make the same kinds of demands on students in each of the four skill/understanding areas, we assign a single-digit measure of the prominence of that skill or understanding area in the problem according to the following scale of weighting codes. *Note that these numbers are not measures of student performance, but are measures of the demands of the task for a given performance action. You might think of this as akin to an Olympic dive or figure skating maneuver being given a numerical rating.*

WEIGHTING CODES:
0 = not present at all
1 = present in small measure
2 = present in moderate measure, and affects solution
3 = a prominent presence
4 = a dominant presence

USING THE TASKS

These assessments can be used in a variety of ways, depending on your local needs and circumstances.

- At each grade level, the balanced assessment tasks provide opportunity for integrated, classroom-based formative assessment. The collection allows you to select tasks that are appropriate at particular points in the curriculum or that specifically address a mathematics action that students need help with. As previously mentioned, the collection is balanced as to both content (math objects) and process (math actions), and is developmentally appropriate for the particular age group. Using the tasks as formative assessment enables the teacher not only to adjust instructional strategy for the whole class, but also to pinpoint individual weaknesses.
- At each grade level, the balanced assessment tasks can be used as exemplars for open-response questions on high-stakes tests. They provide students with the opportunity to work on organizing their mathematical thinking and to practice and refine their communication skills.
- At each grade level, the balanced assessment tasks can serve as a transition toward a standards-based curriculum or as enrichment for existing curriculum. They are designed to be used as a supplement to any standards-based curriculum.
- At each grade level, the balanced assessment tasks can be used as pretest/posttest items for diagnostic purposes. They may also be used as summative assessment for specific topics or for additional information when assigning a mathematics grade.

It is extremely important that you work through a task yourself before giving it to your students. Only in this way can you become familiar with the context and the mathematical demands and be able to anticipate what needs to be highlighted as you launch the task. At the beginning of each task, you will find a teacher's guide that details

- the math object category (some tasks fall into more than one category; the dominant object is indicated by a bold **X**);
- the process, or math actions, weightings;
- assumed mathematical background;
- core elements of performance ;
- specific directions for launching and conducting the task;
- possible extensions;
- materials (paper and pencil are assumed to be necessary for all tasks; in some cases, additional materials and calculators are stipulated).

A pre-activity is provided for many of the tasks. It is expected that the teacher will do the pre-activity with the whole class and answer any questions before assigning the main body of the task.

Depending on whether the task is being used as formative or summative assessment, the launch of a task will vary. In some cases, it will be necessary only to distribute the task to students and let them read and work through it on their own. In other cases, it may be more productive to have them work in pairs, but report back individually. When students are meeting this type of task for the first time, especially when the tasks are being used primarily as learning tasks to enhance the curriculum, you may decide to work through the tasks item by item, talking with the students and posing questions when things get "stuck." This type of informal assessment gives you the opportunity to observe what strategies students favor, what kinds of questions they ask, what they seem to understand and what they are struggling with, and what kinds of prompts get them "unstuck."

Younger students will need assistance in the scribing of their answers—it is very common for verbal exposition to develop before writing ability. By the middle of first grade, however, students should be encouraged to put pencil to paper in order to chronicle their mathematical activity.

We cannot stress enough the importance of teachers working through the entire task completely before using it in the classroom; it is only in this way that teachers can anticipate where their students may run into difficulty. It is also imperative that teachers are aware of, and are comfortable with, all possible solutions—in other words, there is often more than one "right" answer or approach.

USING THE RUBRICS

Rubrics are a set of rules or guidelines for giving scores to student work; they answer the question "What do the varying degrees of mastery for this task look like?" The rubrics that accompany these assessments are based on the Core Elements of Performance that are identified for each task and may be used in a variety of ways. If the tasks are being used as formative assessment, students should be allowed to revise their work to meet as closely as possible the criteria for "full competency." If the tasks are being used as summative assessment, the partial- and full-competency descriptors can be restated as a four-level holistic rubric. Level 4 work meets all the descriptors for "full competency"; Levels 1–3 are arrived at by appropriately adjusting the descriptors for "partial competency."

Teachers may need to translate scores from balanced assessment tasks to a letter- or number-grade system. While it may not be possible to preserve all four aspects of the mathematics actions, these scores should be aggregated no further than a separate score for skills (Transforming/Manipulating) and a separate score for understanding, which combines both Modeling and Inferring. Collapsing the individual scores substantially reduces the utility of these materials to provide a mathematical profile of student understanding and to contribute to making informed decisions about students. Information on using these rubrics as part of a complete scoring system with accompanying software can be found in the document *MCAPS: Mathematics Content and Process Scoring* (Schwartz & Kenney, 1999).

These rubrics can also be used to facilitate professional development activities for teachers, either in a study group that focuses on scoring student work or an action-research project on formative assessment. The conversations that ensue when teachers look at their own students' work, compare it against the criteria set forth in the rubric, and then discuss it with their colleagues, provide a clear perspective as to where students had difficulties and where they were successful. This, in turn, leads to better pedagogy and enhanced mathematical understanding for both teacher and student.

SUMMARY

Any assessment that is truly worthwhile to teachers, students, and others with a valid interest in what students can do mathematically must have the following characteristics, as indicated in NCTM's (1995) *Assessment Standards*:

- **The assessment must focus on important, grade-level-appropriate mathematics.** Since it is not possible to assess everything that students have learned, it is important to select carefully what learning is assessed by concentrating on the most important and useful mathematics taught and learned at that grade level.
- **The assessments must be worthwhile learning activities—not digressions from learning.** For the student, assessment is a tool that helps further the understanding of important mathematical ideas. For the teacher, assessment is student work that informs and augments instruction. Worthwhile assessment is not something students and teachers "stop and do," but a way to further what they are already doing.
- **The assessments must maintain a focus on accessibility and equity for all students.** The student must have, and the teacher and student must perceive that the student has, a fair opportunity to do his or her best. Assessments are designed to provide a student of either gender and of any cultural, linguistic, and socio-

economic background with the means to do his or her strongest mathematical work.
- **The assessments must elicit scorable, informative student work.** The assessments are designed to result in more than just an answer from the student. Rather, students are asked to solve a problem, show their thinking, and create a product. The information in the student's response, and the features of the student's work that are evaluated, give a picture of his or her understanding of mathematical concepts, strategies, tools, and procedures.

It is the hope and intent of the authors that the balanced assessment tasks and rubrics provided in the following chapters will assist teachers in their efforts to put into practice the theoretical guidelines provided by NCTM for the assessment of learning in mathematics. They are designed with great care to make them as revealing and adaptable as possible, suitable for incorporation into any curriculum, and a source of important information about students' mathematical understanding. We hope that teachers will also find within these assessments new ways to think about teaching mathematics, new ways to sustain dialogue with students, and new impetus for conversation with colleagues about the important work in which teachers engage.

2

Number and Quantity

Number and quantity are the basic building blocks of a mathematical system. For this reason, it is essential that these concepts be presented in a coherent, connected way from the earliest days of kindergarten. Since most of the time spent in mathematics classrooms at the primary and elementary level is appropriately focused on the development of number sense, the tasks in this collection that are concerned with the mathematical objects of number and quantity make up 50 percent of the total.

During the primary grades (K–2), students should grow in their ability to:

- demonstrate basic understanding of our numeration system, and answer questions such as *If a rabbit hops three spaces forward on a number line, and then two spaces backward, what number will the rabbit end up on?*
- show evidence of the conceptual meaning of addition and subtraction of whole numbers and integers, and answer questions such as Rachel has 3 carrots and Peter gave her 2 more. *How many carrots does Rachel have now?*
- make reasonable approximations, using standard and nonstandard measures, and answer questions such as *How many carrots tall am I? How do you know?*
- demonstrate basic understanding of the meaning of simple fractional parts, and answer questions such as *Which is bigger, $\frac{1}{4}$ or $\frac{1}{2}$? Why?*
- identify and measure continuous quantity such as length, area, and time, and answer questions such as *How many squares of seed would it take to cover this garden?*

As students move through the elementary grades (3–5) and transition into middle school, we expect them to:

- demonstrate a robust understanding of the conceptual meaning of addition and subtraction of whole numbers and integers, evidenced by the ability to answer questions such as *Sarah has 3 apples and Joe gave her 2 more. How many apples does Sarah have now? Sarah has 3 apples and Joe has 2 more apples than Sarah. How many apples do they have altogether?*
- demonstrate a growing understanding of the various meanings of both multiplication and division of whole numbers and integers, evidenced by the ability to answer questions such as *At a party, 20 bags of candy were given out. Each bag contained 5 candies. How many candies were given out altogether?*
- demonstrate a reasonable degree of computational facility with the four arithmetic operations on whole numbers and integers, evidenced by the ability to answer questions such as *If you add a total of 12 to a given number, what are two subtractions that will bring you back to the original number? If you divide a number by 3 and then by 2, what will you have to multiply the result by in order to get back to the original number?*
- demonstrate ability to make reasonable approximations for the results of arithmetic computations (an expectation still unrealized in most United States elementary classrooms), and to answer questions such as *To the nearest hundred, what is 38 times 42? To the nearest hundred, what is 716 plus 879?*
- demonstrate a growing understanding of the order properties of decimals and other rational fractions, and respond to directives such as *Write a fraction that is larger than ⅓ and smaller than ½. Write a decimal that is larger than .083 and smaller than 0.15.*
- demonstrate ability to identify and measure continuous quantities such as length, area, weight, and time; to make reasonable estimates of lengths, areas, weights, and time in one's environment; and to answer questions such as *How much does a gallon of milk weigh? How do you know? How much time does it take you to say your name? How do you know?*

The tasks in this chapter provide the opportunity to demonstrate these abilities in a variety of ways. Tasks that focus on number and quantity tend to be more heavily weighted toward manipulation and transformation, but they also demand differing amounts of modeling and inference. Communication demands vary from task to task. By choosing tasks that are more heavily weighted in any one of the process areas, teachers are able to individualize the assessment and respond to the varying competencies and weaknesses of each student.

ADD-RINGS

TEACHER'S GUIDE

Grade Level: Primary

Description:

The intent of this task is to have students demonstrate arithmetic computation skills and the ability to recognize a pattern, in particular the relation between addition and its inverse.

Mathematics:

Math Objects

[X] Number/Quantity	[ ] Shape/Space	[X] Function/Pattern
[ ] Chance/Data	[ ] Arrangement	

Math Actions (possible weights: 0 through 4)

[0] Modeling/Formulating	[2] Transforming/Manipulating
[2] Drawing Conclusions	[2] Communicating

Assumed Mathematical Background:

This task assumes very basic addition and subtraction skills.

Core Elements of Performance:

- Arrive at a correct result for all computations, regardless of the initial number chosen and its position in the ring.
- Be able to exhibit an understanding of the arithmetic generalization described by the ring and the inverse relationship of the arithmetic processes involved.

Using This Task:

Read through the prompt with your students to ensure that they understand the task. For informal classroom use, you can adjust the difficulty level by providing more background information and direction to the students, particularly on question 4. You may wish to suggest what number students should choose to start with in question 4, in order to avoid a negative answer. In all likelihood, kindergarten and first grade students will need to give their answers to question 5 orally. Carefully document how the student expresses the pattern or the perceived relationship.

Younger students may benefit from using manipulatives, or their own bodies, to model the ring.

Extension:

Stronger primary-grade students can be encouraged to create their own ring using different additions and subtractions from those provided in the task.

Name: _____ **Date:** _____

Add-Rings

Here is a special kind of number ring. Any number that goes into one of the circles on the ring gets changed around as it goes through the other circles.

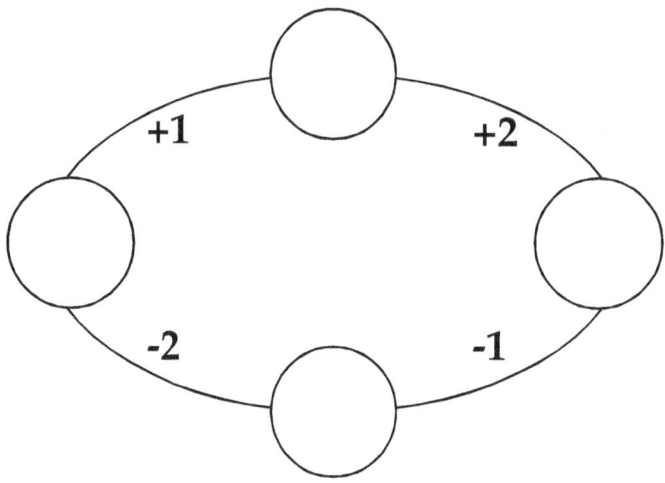

1. Try this ring. Write the number 2 into the top circle and go around the ring. Follow the instructions to add or subtract, and write the correct answer inside each circle. What is your answer when you get back to the top circle?

Name: _____ **Date:** _____

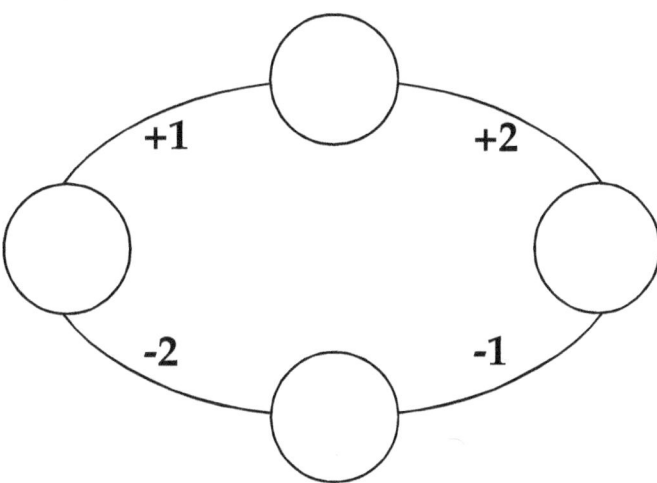

2. Do this again: Put the number 2 into the top circle, but this time go in the opposite direction around the ring. What is your answer when you get back to the top circle?

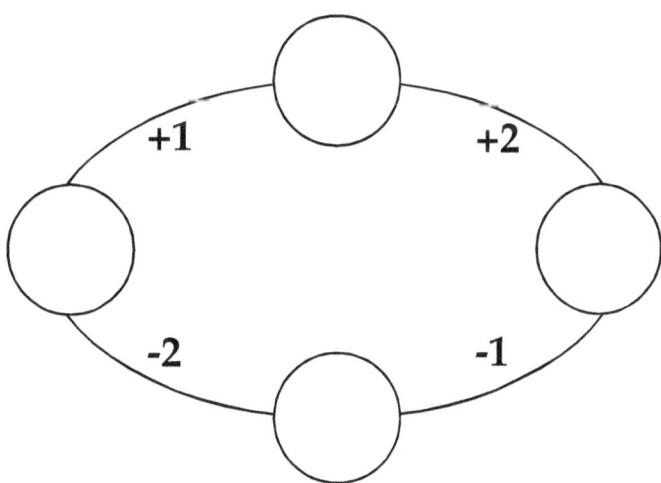

3. Try the ring again. This time put the number 3 in the bottom circle. Go around the ring, following the directions for addition and subtraction, and write the correct answer inside each circle. What is your answer when you get back to the bottom circle?

Name: _____   Date: _____

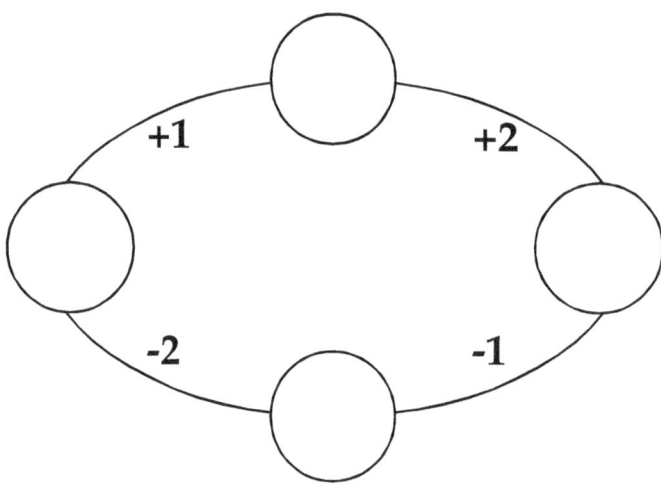

4. Now pick any number you wish and put it in either one of the side circles. Follow around the ring in whichever direction you choose, adding and subtracting as you go. What happens when you get back to the circle where you started?

5. Explain in words why you think you got the answers that you did each time you came back to the circle where you had started.

Add-Rings
SOLUTION AND RUBRIC

1. Regardless of which direction students go around the ring, the computation should lead them to a final answer of 2 when they get back to the top circle.
2. Again, the result will be 2.
3. The resulting answer will be 3. (Make sure students have put the number in the **bottom** circle.)
4. Regardless of the number chosen, the side circle started with, and the direction followed, the final answer will be the same number as the student started with. If younger students have difficulty because an intervening calculation is a zero or a negative number, they may be prompted to use a larger number. Older students, however, should be able to compensate in this event.

Students should in some way indicate that they see the inverse properties of addition and subtraction. They may use terms like "they balance out" or "they add up to zero."

	Partial Competency	Full Competency
Modeling/ Formulating *(weight: 0)*		
Transforming/ Manipulating *(weight: 2)*	Student arrives at a correct result for some of the computations.	Student arrives at a correct result for all of the computations, regardless of the initial number chosen and its position in the ring. If one of the additions or subtractions results in a zero answer or a negative number, the student is able to adjust their strategy.
Inferring/Drawing Conclusions *(weight: 2)*	Student shows some understanding of the inverse relationship that is operating in the ring.	Student demonstrates an understanding of the arithmetic generalization described by the ring, and the inverse relationship of the processes involved.
Communicating *(weight: 2)*	Student provides a fragmentary, incoherent, or incomplete response to question **5**.	Student gives a clear, complete explanation of how the ring works.

Birthday Cupcakes

TEACHER'S GUIDE

Grade Level: Primary

Description:

This task introduces students to the notion of division with remainder through a real-world situation. Formal instruction in division is not necessary.

Mathematics:

Math Objects

- [X] Number/Quantity
- [ ] Shape/Space
- [ ] Function/Pattern
- [ ] Chance/Data
- [ ] Arrangement

Math Actions (possible weights: 0 through 4)

- [1] Modeling/Formulating
- [2] Transforming/Manipulating
- [2] Drawing Conclusions
- [2] Communicating

Assumed Mathematical Background:

This task assumes basic understanding of equivalence.

Core Elements of Performance:

- Divide items evenly among a group.
- Identify the remainder.

Using This Task:

Read through the prompt with your students to ensure that they understand the task. In an informal classroom situation, it may be useful to emphasize to students that full boxes of cupcakes must be purchased.

Extension:

For all but the most able students, question 3 will be a challenging extension.

Materials Needed:

Chips, tiles, or similar manipulatives will help students visualize the process.

Name: _____   Date: _____

BIRTHDAY CUPCAKES

Mr. Ramon wants to buy cupcakes for his son Jaime's birthday party at school. The cupcakes are sold in boxes; each box has either 6 chocolate cupcakes or 6 vanilla cupcakes. There are 27 students in Jaime's class.

1. What is the **least** number of boxes that Mr. Ramon should buy, if each person will get only one cupcake? Show how you find your answer.

2. Will there be any extra cupcakes? If so, how many?

Name: _____   **Date:** _____

Extension:

3. What is the least number of boxes that Mr. Ramon should buy, if each person will get one chocolate **and** one vanilla cupcake? Tell how you get your answer.

Birthday Cupcakes — SOLUTION AND RUBRIC

1. The least number of boxes that Mr. Ramon should buy is 5, since 4 boxes times 6 cupcakes is only 24, not enough for the 27 students in the class. The flavor of the cupcakes is immaterial.
2. If Mr. Ramon buys 5 boxes, there will be 3 extra cupcakes (30 – 27).

Extension:

Using the information from question 1, he should buy twice as many boxes, or ten boxes, in order to give each student twice as many cupcakes. Some students may realize that Mr. Ramon would only need nine boxes if some students were willing to get two cupcakes of the same flavor.

	Partial Competency	Full Competency
Modeling/ Formulating *(weight: 1)*	Student formulates a strategy with which to represent some of the given information.	Student formulates a strategy which takes into account all of the given information.
Transforming/ Manipulating *(weight: 2)*	Student gets a correct numerical answer for either question **1** or **2**.	Student gets correct numerical answers for both questions **1** and **2**.
Inferring/Drawing Conclusions *(weight: 2)*	Student is able to demonstrate an understanding of part/whole in order to infer the least number of boxes needed, but not the remainder.	Student is able to demonstrate an understanding of part/whole in order to infer **both** the least number of boxes and the extra cupcakes.
Communicating *(weight: 2)*	Student gives a partial, unclear, or incomplete explanation for questions **1** and **2**. If a numerical description is used for question **1**, it may be out of sequence or missing a step.	Student gives a full, clear, complete explanation for questions **1** and **2**, utilizing both prose and numerical evidence.

TV Shows

TEACHER'S GUIDE

Grade Level: Primary

Description:

This task asks students to find the duration of several shows on a TV schedule and to make up an ending time for a show that meets certain requirements. All times are in quarter-hour increments.

Mathematics:

Math Objects

[X] Number/Quantity   [ ] Shape/Space   [ ] Function/Pattern

[ ] Chance/Data   [ ] Arrangement

Math Actions (possible weights: 0 through 4)

[1] Modeling/Formulating   [2] Transforming/Manipulating

[2] Drawing Conclusions   [1] Communicating

Assumed Mathematical Background:

This task assumes instruction in measuring time, including units of measure and the concept of elapsed time.

Core Elements of Performance:

- Determine elapsed time from provided information.
- Determine a time that is between two given values.

Using This Task:

Read through the prompt with your students to ensure that they understand the task. This task is designed primarily for second grade use. In an informal classroom situation, it is fruitful to compare and discuss the possibility of multiple correct answers for question 2.

Materials Needed:

It is helpful to have a manipulative analog clock.

Name: _____ **Date:** _____

TV Shows

Here is part of the schedule for TV Channel 37.

4:00 – 4:30 p.m.	Dora the Explorer
4:30 – 5:30 p.m.	Dragon Tales
5:30 – 6:00 p.m.	Bob the Builder
6:00 – 6:15 p.m.	Channel 37 News

1. How many minutes long is each TV show?

    Dora the Explorer:

    Dragon Tales:

    Bob the Builder:

    Channel 37 News:

2. Your favorite TV show starts at 7:00 p.m. on Channel 37.
   It is **longer** than *Bob the Builder* but **shorter** than *Dragon Tales*.
   Make up a possible ending time for this show.

27

TV Shows — Solution and Rubric

1. The shows are respectively 30, 60, 30, and 15 minutes long.

2. The length of the show must be more than 30 minutes but less than 60 minutes. Therefore, the ending time must be later than 7:30 p.m. but earlier than 8:00 p.m. For example, the show could end at 7:45 p.m., which would make the show 45 minutes long.

	Partial Competency	Full Competency
Modeling/ Formulating (weight: 1)	For problem **1**, the student chooses methods for finding time durations that work for only some of the given TV programs (for example, a method that does not work for 4:30–5:30 because it starts and ends in different hours).	For problem **1**, the student chooses methods for finding time durations that work for all of the given TV programs.
Transforming/ Manipulating (weight: 2)	Only some of the time computations are performed correctly.	All necessary time computations are performed correctly.
Inferring/Drawing Conclusions (weight: 2)	For problem **2**, the student's chosen ending time meets only one of the two requirements of the problem ("longer than" and "shorter than"). For example, an answer of "8:00 p.m." would meet the "longer than" requirement, but not the "shorter than" requirement.	For problem **2**, the student's chosen ending time meets both of the requirements of the problem ("longer than" and "shorter than").
Communicating (weight: 1)	Only some numerical answers are clearly written.	All numerical answers are clearly written.

A Very Long Hallway

TEACHER'S GUIDE

Grade Level: Primary

Description:

A hallway of successively numbered rooms acts as a real-world model of the number line. Questions about the distance between rooms can be answered by counting or subtraction.

Mathematics:

Math Objects

[X] Number/Quantity [ ] Shape/Space [ ] Function/Pattern

[ ] Chance/Data [ ] Arrangement

Math Actions (possible weights: 0 through 4)

[2] Modeling/Formulating [2] Transforming/Manipulating

[0] Drawing Conclusions [2] Communicating

Assumed Mathematical Background:

Students should have worked with a number line.

Core Elements of Performance:

- Determine the distance between two objects by counting spaces.
- Use subtraction to find distances when counting is not feasible.

Using This Task:

Read through the prompt with your students to ensure that they understand the task. Kindergarten students should be able to do questions 1 and 2; older students will enjoy the challenge of questions 5–7.

In an informal classroom setting, you may wish to remind students to show how they got their answers.

Extension:

The extensions provided in questions 8–10 are appropriate for advanced Grade 2 students.

Name: _____ Date: _____

A Very Long Hallway

The doors along one side of a very long hallway are numbered in order. The space between one door and the next door is always the same. Here is a picture of part of the hallway.

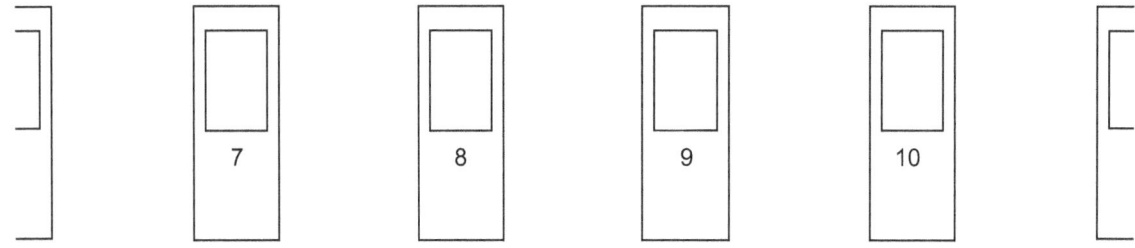

Here are some examples of how the spaces between doors can be counted:

- Door 7 and Door 8 are 1 **space** apart.

- Door 7 and Door 10 are 3 **spaces** apart.

1. How many spaces apart are Door 8 and Door 10?

2. How many spaces apart are Door 2 and Door 9?

3. What are the numbers of two doors that are 5 spaces apart?

Name: _____ **Date:** _____

4. What are the numbers of two other doors that are 5 spaces apart?

5. How many spaces apart are Door 8 and Door 21? Show how you get your answer.

6. What are the numbers of two doors that are 19 spaces apart? Show how you get your answer.

7. How many spaces apart are Door 8 and Door 93? Show how you get your answer.

Extension:

8. What are the numbers of two doors that are 99 spaces apart? Show how you get your answer.

9. What about 9,999 spaces apart? 99,999 spaces apart? Do you see a pattern in your answers? What is it?

A Very Long Hallway

SOLUTION AND RUBRIC

While some of the earlier problems involving small numbers can be done by drawing pictures and counting, it is not practical to draw pictures when dealing with larger numbers. To do these problems efficiently, the student must recognize that the number of spaces between any two doors can be found by subtracting the door numbers.

1. $10 - 8 = 2$, so the doors are 2 spaces apart.

2. $9 - 2 = 7$, so the doors are 7 spaces apart.

3–4. Any pair of doors whose numbers differ by 5 will suffice, such as Door 1 and Door 6, or Door 70 and Door 75.

5. $21 - 8 = 13$, so the doors are 13 spaces apart.

6. Any pair of doors whose numbers differ by 19 will suffice, such as Door 1 and Door 20. Note that this problem requires two-digit subtraction, usually with carrying.

7. $93 - 8 = 85$, so the doors are 85 spaces apart.

Extension:

8–9. One possible set of answers is that Doors 1 and 100 are 99 spaces apart, Doors 1 and 1,000 are 999 spaces apart, Doors 1 and 10,000 are 9,999 spaces apart, Doors 1 and 100,000 are 99,999 spaces apart, and so on.

	Partial Competency	**Full Competency**
Modeling/ Formulating *(weight: 2)*	Student correctly answers the easier questions by drawing and counting, but does not use subtraction at all or uses it incorrectly (for example, giving answers that are consistently off by 1).	Student recognizes how subtraction can be correctly used to determine how many spaces apart two doors are.
Transforming/ Manipulating *(weight: 2)*	Some errors are made in arithmetic computations.	All necessary arithmetic is performed correctly.
Inferring/Drawing Conclusions *(weight: 0)*		
Communicating *(weight: 2)*	For problems that state "Show how you get your answer," the student does not always clearly display the arithmetic computation or other procedure used to answer the question.	For problems that state "Show how you get your answer," the student clearly displays the arithmetic computation or other procedure used to answer the question.

NUMBER AND QUANTITY 33

Dot-to-Dot

TEACHER'S GUIDE

Grade Level: Primary

Description:

This task is designed to assess student understanding of odd and even numbers, and their ability to describe geometric shapes using appropriate vocabulary.

Mathematics:

Math Objects

| [X] Number/Quantity | [ ] Shape/Space | [ ] Function/Pattern |
| [ ] Chance/Data | [ ] Arrangement | |

Math Actions (possible weights: 0 through 4)

| [0] Modeling/Formulating | [2] Transforming/Manipulating |
| [1] Drawing Conclusions | [2] Communicating |

Assumed Mathematical Background:

This task assumes basic understanding of counting numbers, and the ability to skip-count by threes.

Core Elements of Performance:

- Identify sequential even numbers.
- Identify sequential multiples of three.
- Describe geometric shapes.

Using This Task:

Read through the prompt with your students to ensure that they understand the task. In an informal classroom situation you may wish to remind students to fully describe the pictures they have drawn, using appropriate geometry vocabulary.

Extensions:

Question 2 should be considered an extension for kindergarten and early first graders.

Dot-to-Dot

1. Connect in numerical order the dots that have an **even** number.

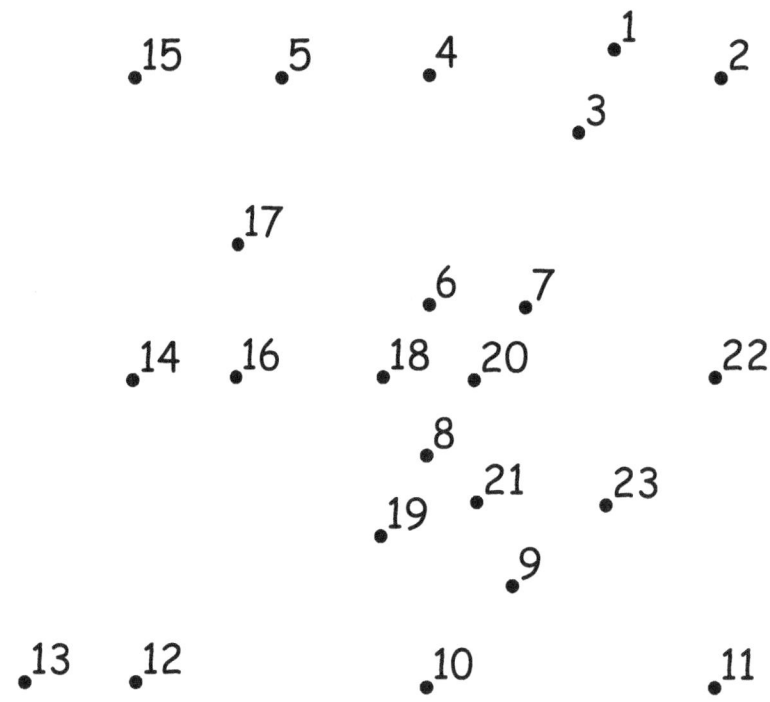

When you are finished, draw a line from the last even number to the first even number.

2. Describe the picture you have drawn.

Name: _____   Date: _____

3. Connect in numerical order the dots that have the numbers you say when you count by 3's, starting with 3. (We have gotten you started.)

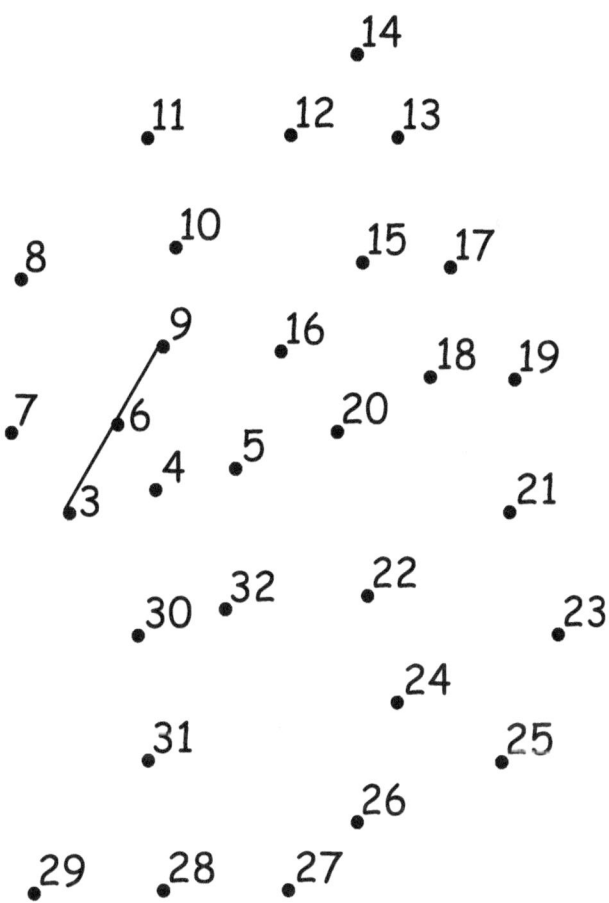

When you are finished, draw a line back to 3, and also a line from 3 to 21.

4. Describe the picture you have drawn.

Dot-to-Dot   SOLUTION AND RUBRIC

1.

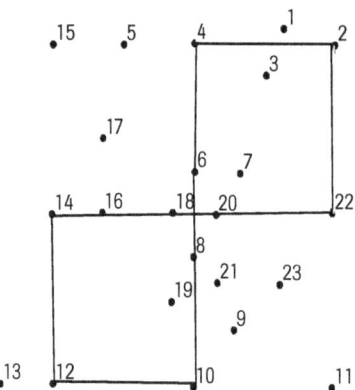

2. Younger students will usually describe this as "two squares hitched together." Older students should be expected to be specific in describing that they are joined at one corner.

3.

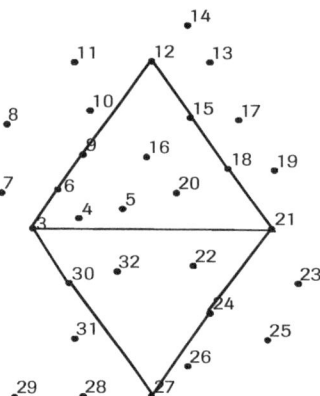

4. Younger students should be expected to describe this as "two triangles hitched together." Older students should be expected to indicate the direction, for example "one pointing up and one pointing down," and the fact that they are joined at their bases. Often this is expressed as "they have one side that is the same."

	Partial Competency	**Full Competency**
Modeling/ Formulating *(weight: 0)*		
Transforming/ Manipulating *(weight: 2)*	Student correctly connects the even numbers (question **1**) or the multiples of three (question **3**).	Student correctly connects the even numbers (question **1**) **and** the multiples of three (question **3**), and follows all additional directions.
Inferring/Drawing Conclusions *(weight: 1)*	Student is able to infer one of the two shapes from the completed picture.	Student is able to infer both of the shapes from the completed picture.
Communicating *(weight: 2)*	Student uses limited geometric vocabulary in describing the resulting shapes.	Student uses appropriate geometric vocabulary in describing the resulting shapes.

MEASURING THE MARIGOLDS

TEACHER'S GUIDE

Grade Level: Primary

Description:

This task requires students to make estimations of measurement using atypical units and to utilize this measure in determining area and perimeter of various shapes.

Mathematics:

Math Objects

| X | Number/Quantity | ☐ Shape/Space | ☐ Function/Pattern |
| ☐ | Chance/Data | ☐ Arrangement | |

Math Actions (possible weights: 0 through 4)

- [2] Modeling/Formulating
- [2] Transforming/Manipulating
- [3] Drawing Conclusions
- [1] Communicating

Assumed Mathematical Background:

Students should have some experience with measurement and making estimates. They should also understand the concept of doubling.

Core Elements of Performance:

- Measure objects in non-standard units.
- Generalize and extend results of measurement to new situations.

Using This Task:

Read through the prompt with your students to ensure that they understand the task. Kindergarten students will enjoy question 1; first grade students should be able to complete the task without difficulty. It may be helpful to photocopy multiple inchworms for younger students to cut out and use. This task lends itself to use as a theme for an interdisciplinary unit with art and music.

Extension:

Implicit in and important to this task is the distinction between area as a measurement that involves both length and width, and perimeter as a measurement that is only linear. With older students you may wish to extend and reinforce this concept by asking, in question 3, "How many inchworms would be needed to cover the garden if the worms were half as fat? Twice as fat?" Another interesting topic is whether the "fatness" of the inchworm makes any difference when measuring around the border.

Name: _____   Date: _____

MEASURING THE MARIGOLDS

Inchworm, Inchworm, measuring the marigolds,
You and your arithmetic
Will probably go far.
Inchworm, Inchworm, measuring the marigolds,
Seems to me you'd stop and see
How beautiful they are.

Here's an inchworm you can use for measuring:

1. Ms. Muffet has planted a lovely garden of marigolds. Here is a picture of one of her flowers.

    About how many inchworms tall is this marigold?

Name: _____ **Date:** _____

2. Jack the Giant has also planted marigolds, which are **twice** as tall as Ms. Muffet's.

   About how many inchworms tall are his marigolds?

3. Here is a picture of Ms. Muffet's garden, as seen by a dragonfly. Because the inchworms in her garden are so fat, each square of her garden can hold just four inchworms.

   a. How many inchworms would it take to cover the whole garden?

   b. How many inchworms would it take to make a border all around this garden?

Measuring the Marigolds

SOLUTION AND RUBRIC

1. Regardless of how students use the inchworm (some may measure it with a ruler, others may measure with their fingers, still others may need to have a cutout copy to manipulate), they should come up with the answer that Ms. Muffet's marigold is about **4** inchworms tall.

2. Some students may get an answer of **8** inchworms tall by adding 4 and 4; others may multiply 4 times 2.

3. Younger students may need to actually place four inchworms in each square of the garden and count to **24**. Others will think of it as an addition problem: 4 and 4 are 8, 8 and 8 are 16, 16 and 8 are 24. Others will do it as multiplication: 6 squares times 4 inchworms/square is 24.

4. Here again, there are a variety of correct approaches to get the answer of **10**. Younger students may need to place, draw, and count. Others will realize that each square is one inchworm on a side, so counting up the outside edges leads to 10. Others may group it as 2 + 3 + 2 + 3 = 10.

	Partial Competency	**Full Competency**
Modeling/ Formulating *(weight: 2)*	Be able to use the inchworm as a unit of measure with some success in questions **1** and **3**.	Be able to use the inchworm as a unit of measure with total success in questions **1** and **3**.
Transforming/ Manipulating *(weight: 2)*	Arrive at a correct numerical answer for one or two of the questions.	Arrive at a correct numerical answer for questions **1**, **2**, and **3**.
Inferring/Drawing Conclusions *(weight: 3)*	Either use the answer from question **1** to correctly answer question **2**, or use the given information to answer one of the two parts of question **3**.	Use the given information and the answer to question **1** in an efficient way to correctly answer questions **2** and **3**.
Communicating *(weight: 1)*	Give most answers in clear numerical or prose form.	Give all answers in clear numerical or prose form.

Table Talk

Grade Level: Primary

TEACHER'S GUIDE

Description:

Students are asked to construct, identify patterns in, and find properties of a 1-to-9 addition table.

Mathematics:

Math Objects

- [X] Number/Quantity
- [ ] Shape/Space
- [ ] Function/Pattern
- [ ] Chance/Data
- [ ] Arrangement

Math Actions (possible weights: 0 through 4)

- [2] Modeling/Formulating
- [2] Transforming/Manipulating
- [3] Drawing Conclusions
- [2] Communicating

Assumed Mathematical Background:

Students should have worked with sequential numbers.

Core Elements of Performance:

- Use the given table to determine how the numbers inside the table are found from the outside numbers.
- Explain in words how the table works.
- Use the table to perform the required additions.

Using This Task:

Read through the prompt with your students to ensure that they understand the task. Be sure that the students understand what is meant by the "black" numbers and the "gray" numbers. The reading level of this task is fairly high; younger students will need to have the prompt read to them. The task should not be administered in one sitting; questions 1–3 make a good class time exercise, questions 4 and 5 may also be paired, question 6 might be a homework assignment.

Extension:

Question 4–6 will be an extension for most Grade 1 students.

TABLE TALK

In this table, the numbers on the top (gray numbers) and the numbers on the side (black numbers) can tell you what number goes in each square.

	1	2	3	4	5	6	7	8	9
1	2	3	4	5	6	7	8	9	10
2	3	4	5	6	7	8	9		
3	4	5	6	7	8				
4	5	6	7	8	9				
5	6								
6									
7									
8									
9									18

Name: _____ **Date:** _____

2. Look carefully at the numbers that are already in the table and figure out a rule for filling out the table. Fill out the rest of the table using this rule.

   a. What number goes in the square that has top number 7 and side number 8?

   b. If the table kept going, what number would be in the square with top number 30 and side number 20?

   c. How can you find the number that goes in any square if you know the square's top number and side number?

3. If you know the number in a square, can you figure out the square's top number and side number? Tell why or why not.

Name: _____   **Date:** _____

4. **a.** A square contains the number 12. Its side number is **7**. What is its top number?

    b. Again imagine that the table keeps going. A square contains the number 45. Its side number is **13**. What is its top number?

    c. If you know the number in a square and you know the square's side number, tell how to find the square's top number.

    d. If you know the number in a square and you know the square's top number, tell how to find the square's side number.

Name: _____ Date: _____

5. If a square's side number and top number are the same, what is special about the number in the square?

6. Here is another copy of the table. Color in every square that should have an even number. **Do not** color the squares that should have odd numbers.

	1	2	3	4	5	6	7	8	9
1									
2									
3									
4									
5									
6									
7									
8									
9									

Table Talk — SOLUTION AND RUBRIC

1. Each row of the table has consecutive numbers in it. Also, each column has consecutive numbers in it. The table can be filled out simply by increasing the numbers by one as you move down from the first row.

The numbers also represent an addition table from 1 to 9, that is the number in each square is the sum of the row number (black) and the column number (gray).

2. a. The number in a square with top number 7 and side number 8 is 15.

   b. Here it becomes important to notice that the number in each square is the sum of the top number and the side number. The number in the square is 50.

   c. To find the number in the square, it is most efficient to add the top number and the side number. However, it is possible to approach this question in slightly different ways. For example, instead of giving a formulaic definition (top number plus side number), a student might give a procedural definition: for example, one could say that the first number in a row is one more than the side number, or the second number in a row is two more than the row number, etc. Then, for instance, the 20th number in a row, that is the number in a cell with top number 20, is 20 more than the side number. While this definition may appear very different from the formulaic definition, the two approaches are functionally equivalent.

3. The answer, in general, is "no." The same number may appear in several consecutive rows. For example, number 5 appears in the first, second, third, and fourth rows. So, it would not be possible to tell what the side and the top numbers are for a square with 5 in it.

4. a. The number is 5. It can either be computed or found directly from the table.

   b. Here it becomes more difficult, once again, to find the number by manipulating the table directly. The top number is 32.

   c–d. Since the number in the square is the sum of the side number and the top number, if one of these numbers is known, the other is the difference between the number in the square and the known number. It is imperative for students to understand that the process of finding the top or the side number from the number in the square and the other known number is the *inverse* of the process of finding the number in the square from the top and the side numbers.

5. If the top and the side numbers are the same, the number in the square is the sum of two equal numbers, or, simply, a double of either the side or the top number. So, to find the number in the square, simply double one of the given numbers.

6. The table should show a checkerboard pattern.

	Partial Competency	**Full Competency**
Modeling/ Formulating *(weight: 2)*	Student applies procedural rules throughout and does not form abstractions from the rules.	Student is able to state general rules, with only indirect reference to the search process.
Transforming/ Manipulating *(weight: 2)*	Student completes some of the computations correctly and fills out the table according to a specific rule, with a few minor computational errors. Shows only a partial pattern in question **6**.	Student completes all the computations correctly and fills out the table completely according to a single specific rule, even if this rule is not correct. Shows complete checkerboard pattern in **6**.
Inferring/Drawing Conclusions *(weight: 3)*	Student completes the table and computes the values using only the consecutive number patterns.	Student completes the table correctly and draws general conclusions about the rules for finding numbers; also uses these rules to find the numbers *outside* of the given table. Correctly determines the even and odd numbers in question **6**.
Communicating *(weight: 2)*	Student completes some of the table and/or gives incomplete reasoning for other answers.	Student completes the table and answers each question about general rules completely.

Leopard's Leap

TEACHER'S GUIDE

Grade Level: Primary/Elementary

Description:

Students are asked to compare small multiples of two numbers in real-world context.

Mathematics:

Math Objects

- [X] Number/Quantity
- [ ] Shape/Space
- [ ] Function/Pattern
- [ ] Chance/Data
- [ ] Arrangement

Math Actions (possible weights: 0 through 4)

- [2] Modeling/Formulating
- [2] Transforming/Manipulating
- [2] Drawing Conclusions
- [2] Communicating

Assumed Mathematical Background:

Students should have a sense of relative distance.

Core Elements of Performance:

- Demonstrate understanding of the phrase "3 meters further" by correctly determining the length of an antelope leap.
- Use the additive property to determine how far the antelope will leap in three jumps.
- Compare the distance covered by the leopard and the antelope.

Using This Task:

Read through the prompt with your students to ensure that they understand the task.

Extension:

The task is designed to be used at grade two and above. This is a challenging multistep task that requires students to formulate an effective way to calculate and represent the required values.

Name: _____   Date: _____

LEOPARD'S LEAP

The leopard and the antelope are playing a game. A leopard can leap 7 meters in one jump. An antelope can leap 3 meters further in one jump.

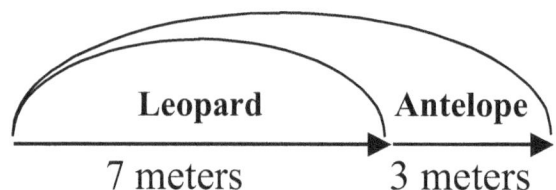

The antelope jumped three times.

How many jumps does the leopard need to make to catch up?

Draw a picture to show how you get your answer.

Leopard's Leap

SOLUTION AND RUBRIC

A second grade student is likely to draw a diagram, similar to the one given in the problem, that includes all three of the antelope's jumps. Each of antelope's jumps is $7 + 3 = 10$ meters, so three jumps measure $10 + 10 + 10 = 30$ meters.

The second part of the diagram might include the leopard jumping to the 7-meter, 14-meter, 21-meter, 28-meter, and 35-meter marks. To catch the antelope, the leopard needs to get slightly further than four jumps (28 meters), so the leopard needs five jumps to catch up.

The problem can also be solved numerically with similar computations or by subtracting 7 meters repeatedly from 30 meters until the fifth subtraction, when the remaining quantity is too small to subtract successfully. That is,

$30 - 7 = 23$ First jump
$23 - 7 = 16$ Second jump
$16 - 7 = 9$ Third jump
$9 - 7 = 2$ Fourth jump
2 meters Fifth jump (not full 7 meters)

If the answer is obtained numerically, a diagram may follow the computations.

	Partial Competency	Full Competency
Modeling/ Formulating *(weight: 2)*	Make a diagram that reflects the conditions of the problem.	Make a diagram that helps the computations or reports the results of the computations graphically.
Transforming/ Manipulating *(weight: 2)*	Complete some of the computations correctly.	Complete all computations correctly.
Inferring/Drawing Conclusions *(weight: 2)*	Conclude that the antelope's leap is 10 meters. Develop a local strategy for finding the number of leopard leaps.	Develop a general strategy for finding the number of leopard leaps. Conclude that at least one of the necessary leaps will be shorter than maximum, but that it must be counted in the total.
Communicating *(weight: 2)*	Report only the answer and partial computations and diagrams.	Provide the diagram, computations, and explanations to show the solution path.

Add 'Em Up

TEACHER'S GUIDE

Grade Level: Primary/Elementary

Description:

This task assesses student ability to use addition skills in an unusual format. Students form equal sums with pairs and triples of different numbers.

Mathematics:

Math Objects

| [X] Number/Quantity | [ ] Shape/Space | [ ] Function/Pattern |
| [ ] Chance/Data | [ ] Arrangement | |

Math Actions (possible weights: 0 through 4)

| [0] Modeling/Formulating | [3] Transforming/Manipulating |
| [2] Drawing Conclusions | [1] Communicating |

Assumed Mathematical Background:

This task assumes basic arithmetic instruction and some notion of equivalency.

Core Elements of Performance:

- Arrive at a correct result for all computations.
- Develop a strategy that will lead to a successful result.

Using This Task:

Read through the prompt with your students to ensure that they understand the task.

Remind students that each number can be used only once. If more intervention is needed, it may be sufficient to suggest that they temporarily forget about the middle circle in question 3 and concentrate on making opposite pairs of circles add to the same sum.

This task provides many challenges for younger students; they may have difficulty adding in a horizontal orientation, and they may be challenged to find an efficient strategy for question 3.

Extension:

The provided extension question is a challenging exploration of a magic square. Students who are successful with question 3 could also be asked to come up with a different solution using another number in the center circle.

Add 'Em Up

1. Can you place the numbers **1, 2, 4,** and **5** in the blank circles, so that each number row adds up to the same total?

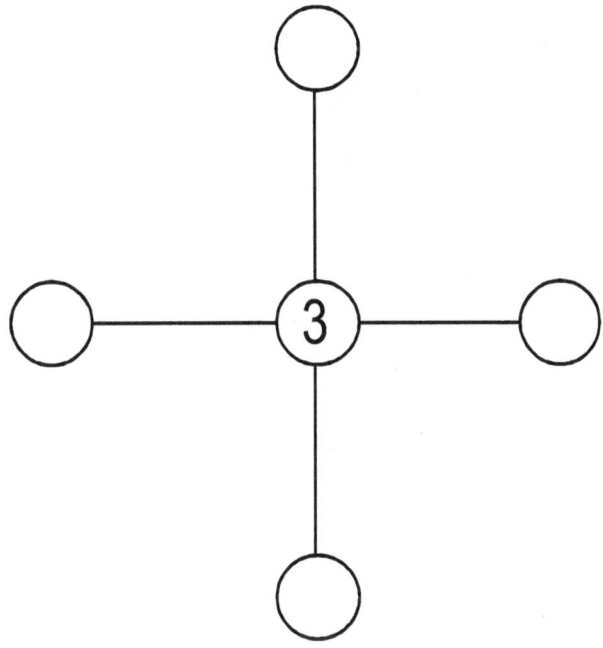

2. On the left are the numbers from **1** to **9**. Use eight of these numbers to fill the blank circles in the Big Wheel, so that when you add any two numbers connected by a line, you get **10**.

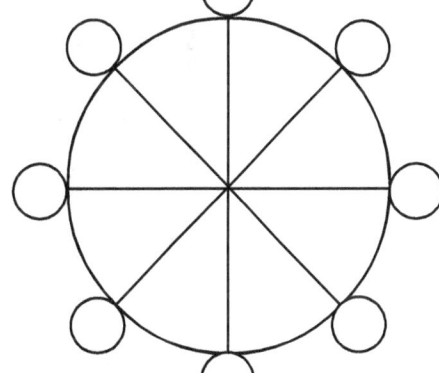

Name: _____ **Date:** _____

3. Here is a different kind of Big Wheel. Can you place each number from **1** through **9** in one of the blank circles, so that each number row adds up to the same total? Remember, you can use each number only once.

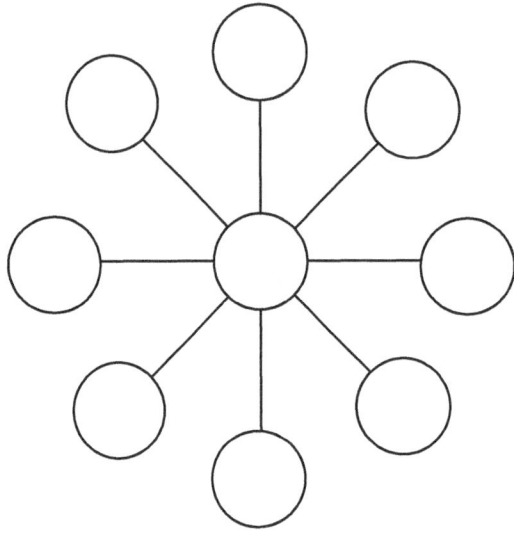

Extension:

4. Place the numbers from 1 to 9 into the boxes of this table, so that every three numbers connected by a dotted arrow add up to the same total.

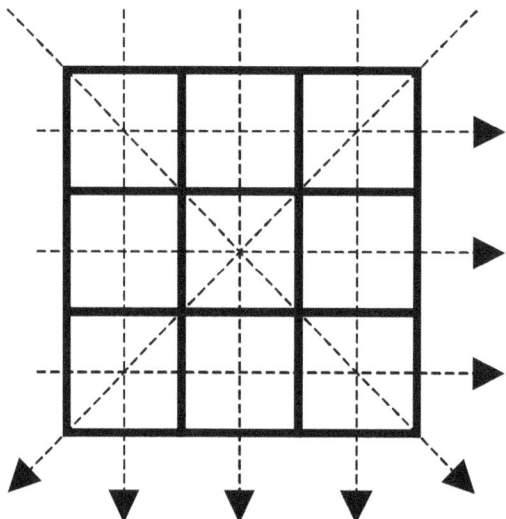

Add 'Em Up

SOLUTION AND RUBRIC

1.

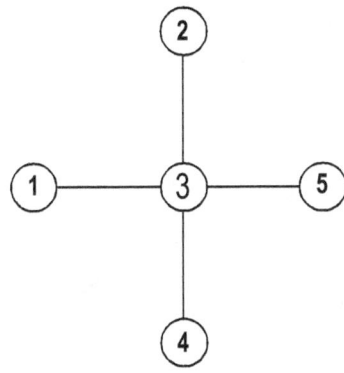

One of the rows must contain the numbers 1 and 5, the other 4 and 2. Obviously it is equally correct to interchange the position of the 2, 4 and the 1, 5 number pairs.

2. There are several ways to solve this problem. From question 1, students should notice that the largest number should be paired with the smallest number, the second largest with the second smallest and so forth. There are three solution possibilities:

The group of smaller numbers is 1, 2, 3, and 4; the corresponding group of larger numbers is 8, 7, 6, and 5; each pair adds to 9.

The group of smaller numbers is 1, 2, 3, and 4; the corresponding group of larger numbers is 9, 8, 7, and 6; each pair adds to 10.

The group of smaller numbers is 2, 3, 4, and 5; the corresponding group of larger numbers Is 9, 8, 7, and 6; each pair adds to 11.

For example:

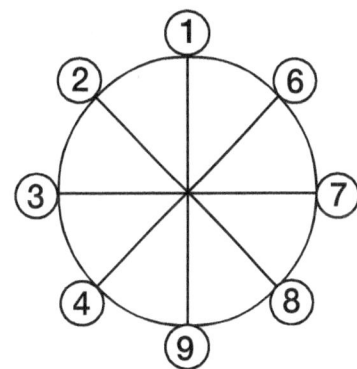

As long as the number pairs are kept intact, it does not matter how they are arranged around the circle.

3. Some students may solve this problem "from scratch"; others may notice that it is isomorphic to question 2.

Here are three possible solutions for this wheel. As long as number pairs that add to 10, 9, and 11 respectively are preserved, they may be arranged in any sequence around the wheel; the corresponding sums in the wheel are 15, 18, and 12.

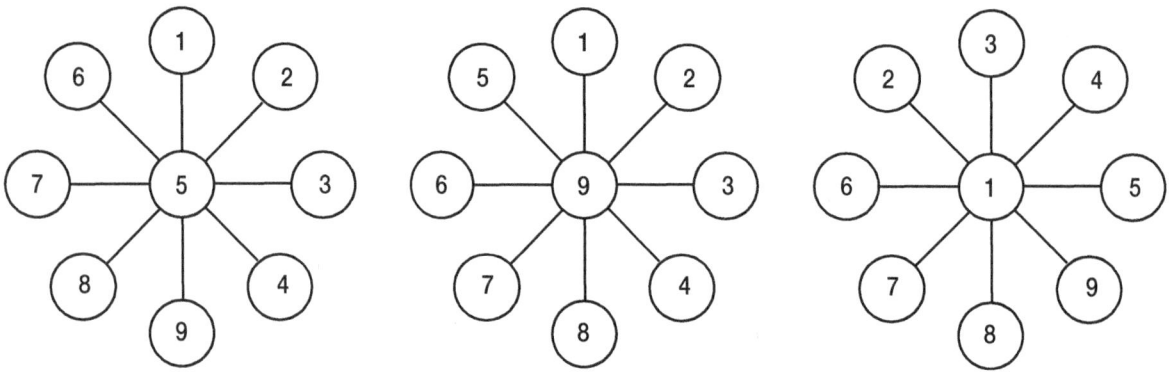

Extension:

4. Students should notice that the arrangement must be similar to one of the arrangements in question 3. With some trial and error, it is easy to discover that the number in the center must be 5. Then 1 and 9 cannot be in the corners—if 9 is in a corner, then 6, 7, and 8 cannot be in the any of the corners. Therefore, 1 and 9 must be in a middle row or column. The only possible row with one in the middle is 6, 1, 8; 6 and 8 are in the corners next to 1, and 4 and 2 are in the corners next to 9. This leaves only two possible positions for 3 and 7, of which only one is acceptable.

The final arrangement must be similar to

2	7	6
9	5	1
4	3	8

Again, the actual arrangement may be a rotation or a reflection of the one given above.

NUMBER AND QUANTITY

	Partial Competency	Full Competency
Modeling/ Formulating *(weight: 0)*		
Transforming/ Manipulating *(weight: 3)*	Student arrives at a correct result for some of the computations.	Student arrives at a correct result for all of the computations and verifies that the sums are equal.
Inferring/Drawing Conclusions *(weight: 2)*	Student develops a strategy that will lead to a successful result in one or two of the questions.	Student develops a strategy that will lead to a successful result in all of the questions, guided by the results from previous calculations.
Communicating *(weight: 1)*	Student places the numbers in the correct position in some of the circles for parts of the problem that have been completed.	Student places the numbers in the correct position in all of the circles.

Multiplication Rings

TEACHER'S GUIDE

Grade Level: Elementary

Description:

The intent of this task is to have students use arithmetic computation skills to identify a pattern involving inverse operations

Mathematics:

Math Objects

[X] Number/Quantity [ ] Shape/Space [ ] Function/Pattern

[ ] Chance/Data [ ] Arrangement

Math Actions (possible weights: 0 through 4)

[2] Modeling/Formulating [2] Transforming/Manipulating

[2] Drawing Conclusions [2] Communicating

Assumed Mathematical Background:

This task assumes basic arithmetic instruction in addition and subtraction.

Core Elements of Performance:

- Arrive at a correct result for all computations, regardless of the initial number chosen and its position in the ring.
- Exhibit an understanding of the arithmetic generalization described by the ring, and the inverse relationship of the processes involved.

Using This Task:

Read through the prompt with your students to ensure that they understand the task. For informal classroom use, whether instructional or diagnostic, you may want to adjust the demand of the task by providing more background information since the original version does not provide any basic information about the inverse relationship of multiplication and division, nor any hints about what initial number to use.

Extension:

Changing the direction of movement around the ring may lead to an intermediate fractional result. You can also have students design a ring with a larger number of multiplication and division commands.

Name: _____   Date: _____

MULTIPLICATION RINGS

Here is a special kind of number ring. Any number you put into one of the circles gets changed as it goes around the ring.

1. Try this ring. Put a number into the top circle and follow around the ring, multiplying and dividing as you go

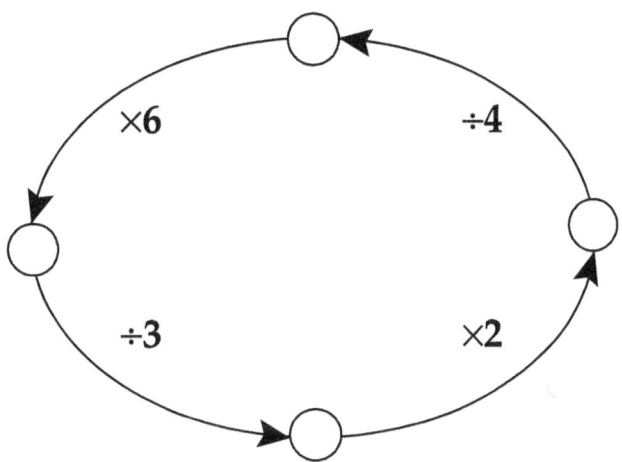

Write your final answer above the number you started with.

Describe in words what happened.

Name: _____ **Date:** _____

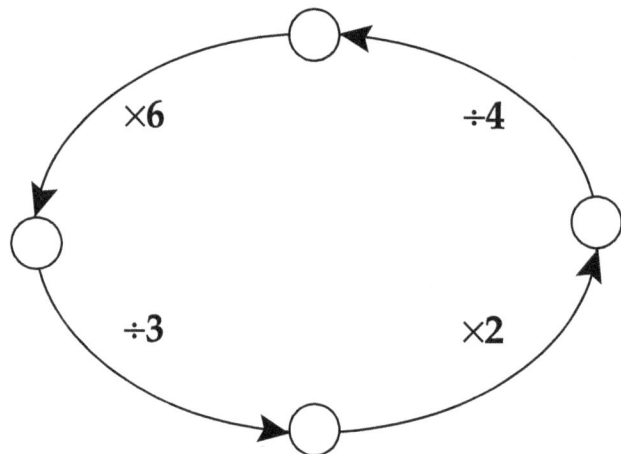

2. Try this again with another number. What is the result? Describe in words what happened.

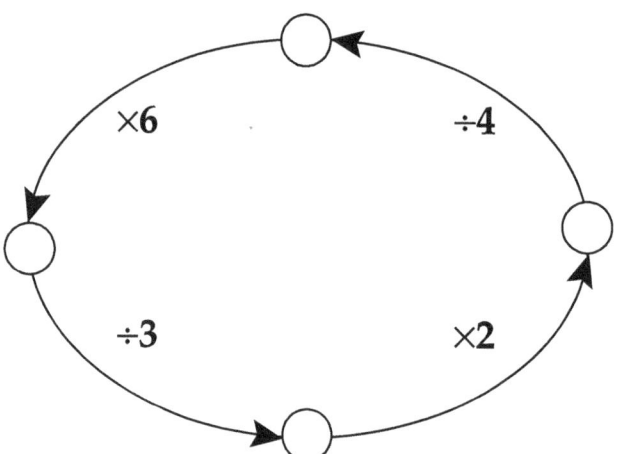

3. Try the ring again, but this time put a number into the **bottom** circle and follow around the ring. What is the result when you return to your starting point?

Name: _____ **Date:** _____

4. Design your own ring, but **use different numbers for the multiplication and division instructions**. Make sure that your ring works out the same way as the original did.

5. Explain in words why you think the ring works the way it does.

Multiplication Rings

SOLUTION AND RUBRIC

No matter what number is put into the top slot in **1** and **2**, going through the other three circles and returning to the top produces the same number. Using a number smaller than 2 will lead to a fractional result in question **3**; students should be able to adjust to this by using a larger number. These results are all due to the fact that the operations are balanced or counteracted by each other—the combined effect of multiplying by 6 and 2 and dividing by 4 and 3 is the same as multiplying by 12, and then dividing by 12.

	Partial Competency	Full Competency
Modeling/ Formulating *(weight: 2)*	In **5**, design a ring that may not "work" in the sense of multiplying and dividing by equivalent amounts, or that uses the same numbers as the original ring in a different order.	In **5**, design a ring that is entirely correct and uses different numbers from the original ring.
Transforming/ Manipulating *(weight: 2)*	Provide correct numerical results for all but one of questions **1–3**.	Consistently get correct numerical results. Be able to adapt to fractions as intermediate results.
Inferring/Drawing Conclusions *(weight: 2)*	Recognize that behavior of ring is independent of starting point. Get the correct answer in **3**, but not be able to articulate the results in **4**. Design a ring in **5** that is little more than a copy of the given ring.	Articulate the conclusion with respect to the initial position in the ring and provide at least a rudimentary generalization of results in **1-3**. Exhibit an understanding of the process of the ring, both through a clear verbal answer to **4** and a ring design in **5** that uses completely different numbers and sequence of multiplication and division.
Communicating *(weight: 2)*	Present evidence in **1–3** by exhibiting some calculations or intervening answers without a clear verbalization of the final result. Give a partial or unclear explanation for **4**.	Give a full, clear explanation for all questions.

NUMBER AND QUANTITY   63

Fermi Four

TEACHER'S GUIDE

Grade Level: Elementary

Description:

This task assesses student ability to make reasonable estimates of quantities with which they are somewhat familiar and to express these estimates in atypical units.

Mathematics:

Math Objects

| [X] Number/Quantity | [ ] Shape/Space | [ ] Function/Pattern |
| [ ] Chance/Data | [ ] Arrangement | |

Math Actions (possible weights: 0 through 4)

| [3] Modeling/Formulating | [3] Transforming/Manipulating |
| [1] Drawing Conclusions | [1] Communicating |

Assumed Mathematical Background:

This task assumes a basic understanding of estimation and also the ability to convert from one measurement system to another.

Core Elements of Performance:

- Make use of information not contained in the statement of the problem.
- Make decisions about the reasonableness of an estimate.
- Convert from one measuring system to another and express quantities in atypical units.

Using This Task:

Read through the prompt with your students to ensure that they understand the task. The pre-activity for this task is essential, as it is almost always necessary to review the process of converting from one measurement system to another and the manipulation of the units. Students should be encouraged to use any manipulatives or reference materials they wish; they should also have calculators available. This task lends itself nicely to being done in pairs or small groups.

Extension:

Students enjoy making up their own Fermi problem once they have had the opportunity to work the provided questions.

Materials Needed:

Calculators, unit conversion chart or fact sheet.

Name: _____ Date: _____

Fermi Four

A famous teacher and physicist named Enrico Fermi liked to ask interesting questions of his students. Think about the questions below, and give your answers in the appropriate units.

Pre-Activity:

Think about the speed limit on the main street of your town. How would you express this in **feet per hour**?

Now think about the speed limit on a major highway. How would you express this in **miles per minute**?

Task:

1. Your parents tell you that if you are good for 1 million seconds you will get a special treat. Will you have to be good for almost 12 hours, or 12 days, or 12 weeks?

2. If you lay pennies side-by-side in a straight line so that they are touching each other, what is the value, **in dollars**, of a mile-long row of pennies?

3. It takes about four times as long to drive from Boston to New York as it does to fly. What do you think is average speed of an airplane? Express this number in **feet per minute**.

FERMI FOUR

SOLUTION AND RUBRIC

Pre-Activity:

The pre-activity is designed to familiarize students with the arithmetic process necessary to convert from one unit to another. For the first example, students are most likely to think of a speed in $\frac{\text{miles}}{\text{hour}}$; they must multiply this quantity by $\frac{5280 \text{ feet}}{1 \text{ mile}}$ to obtain a quantity measured in $\frac{\text{feet}}{\text{hour}}$. For the second problem, the student must divide the chosen quantity in $\frac{\text{miles}}{\text{hour}}$ by $\frac{60 \text{ minutes}}{\text{hour}}$ to obtain a quantity measured in $\frac{\text{miles}}{\text{minute}}$. For all of these problems it is important to understand that multiplying by a factor such as $\frac{5280 \text{ feet}}{1 \text{ mile}}$ does not change a quantity's value—only the units of expression.

Task:

1. A million seconds divided by $\frac{60 \text{ seconds}}{1 \text{ minute}}$, then divided by $\frac{60 \text{ minutes}}{1 \text{ hour}}$, then divided by $\frac{24 \text{ hours}}{1 \text{ day}}$, becomes about 11.6 days.

2. A penny is about ¾ of an inch wide, so we can begin with the ratio $\frac{1 \text{ penny}}{0.75 \text{ inch}}$. Multiplying by $\frac{12 \text{ inches}}{1 \text{ foot}}$, then by $\frac{5280 \text{ feet}}{1 \text{ mile}}$, we get $\frac{12 \cdot 5280 \text{ pennies}}{0.75 \text{ mile}}$ which simplifies to 84,480 $\frac{\text{pennies}}{\text{mile}}$. A mile of pennies is worth about $845!

3. One could estimate the plane's speed to be about 220 $\frac{\text{miles}}{\text{hour}}$ (four times a typical highway speed of 55 $\frac{\text{miles}}{\text{hour}}$). Multiplying by $\frac{5280 \text{ feet}}{1 \text{ mile}}$, then by $\frac{1 \text{ hour}}{60 \text{ minute}}$, gives a speed of 19,360 $\frac{\text{feet}}{\text{second}}$. So, the plane's speed is about 19 to 20 thousand feet/minute.

	Partial Competency	**Full Competency**
Modeling/ Formulating *(weight: 3)*	Identify the conversion information that is needed.	Design a sequence of computations for each problem which yields the desired result.
Transforming/ Manipulating *(weight: 3)*	Perform correct sequences of computations for some of the problems, perhaps making a few errors of inverting factors (e.g., multiplying by 12 instead of 1/12 when converting inches to feet).	Perform accurate computations with no inversions of factors.
Inferring/Drawing Conclusions *(weight: 1)*	Indicate an awareness of the reasonableness of the results in some questions.	Indicate an awareness of the reasonableness of the results in all questions.
Communicating *(weight: 1)*	Communicate all answers clearly, but without indicating units of measure.	Clearly communicate all results, using the required units

Broken Calculators

TEACHER'S GUIDE

Grade Level: Elementary

Description:

Students are asked to demonstrate their understanding of the arithmetic calculation algorithms that overcome deliberately introduced deficits in the standard algorithms.

Mathematics:

Math Objects

- [X] Number/Quantity
- [ ] Shape/Space
- [ ] Function/Pattern
- [ ] Chance/Data
- [ ] Arrangement

Math Actions (possible weights: 0 through 4)

- [0] Modeling/Formulating
- [2] Transforming/Manipulating
- [3] Drawing Conclusions
- [2] Communicating

Assumed Mathematical Background:

This task assumes a basic understanding of the arithmetic calculation algorithms.

Core Elements of Performance:

- Devise alternative arithmetic calculations that overcome given constraints.
- Develop a general approach to solve each problem and be able to verbalize the method used.

Using This Task:

Read through the prompt with your students to ensure that they understand the task.

The pre-activity to this task is very important; some students will need to have the given constraints restated or elaborated upon.

The task can be done without calculators, but for most students it is beneficial to actually have one in hand to model the constraints and to test out alternative solutions. Students should be reminded to write out all the mental arithmetic and the calculator steps they employ in their solution.

Extension:

This task may be extended by imposing different constraints and by requiring students to find the most efficient alternative.

Materials Needed:

Calculators

Name: _____   Date: _____

BROKEN CALCULATORS

Pre-Activity:

Suppose your calculator is broken so that only the 0 and the 1 and the + and the − keys work. How can you get your calculator to show the year in which you were born?

Task:

1. Suppose the + key on your calculator is broken. Show a way to add that doesn't require that key.

2. When they fixed the + key on your calculator at the repair shop, they broke the × key. Show a way to multiply that doesn't require that key.

3. As in the pre-activity, suppose your calculator is broken so that only the 0 and the 1 and the + and the − keys work. What is the smallest number of steps you need to get your calculator to show the year in which you were born?

Broken Calculators

SOLUTION AND RUBRIC

Pre-Activity:

Some of the ways 1984 may be represented are 1000 + 100 nine times + 10 eight times + 1 four times **or** 1111 + 111 three times + 110 four times + 100 **or** 1000 + 1000 − 10 − 10 + 1 + 1 + 1 + 1.

Task:

1. Adding without the addition key will most likely be demonstrated at this grade level by choosing to add a number to itself through multiplication, that is $a + a = 2a$. It may also be accomplished either by thinking of it as $a + b = 0 − [0 − a − b]$ or $a + b = c$, so $b = c − a$, where c is determined by a process of trial and error, but few elementary students will realize this. For example, suppose that we need to add 67 and 84. We know that 70 is more than 67 and 90 is more than 84, so therefore we know that 70 + 90 is more than 67 + 84. We know that 70 is 3 more than 67, and 90 is 6 more than 84. If we add 70 and 90 in our heads to get 160 and then subtract the 3 and the 6, we get 151.

2. Multiplication can be thought of as a process of addition: $a \cdot b = a + a + a + \ldots + a$ (b times). At this grade level, most students will use simple integers less than 10. However, it is possible to use a similar approach with larger numbers. For example, suppose we want to multiply 46 by 27. This means we want the value of $(40 + 6) \times (20 + 7)$, so we need to know the following amounts: How many are forty twenties? (40×20) How many are forty sevens? (40×7) How many are six twenties? (6×20) How many are six sevens? (6×7) To calculate these amounts, all we need to know are the multiplication tables (mental arithmetic) and how to deal with powers of 10. The required values are $800 + 280 + 120 + 42 = 1242$.

	Partial Competency	Full Competency
Modeling/ Formulating *(weight: 0)*		
Transforming/ Manipulating *(weight: 2)*	Make no more than one computational error per question.	Make no computational errors.
Inferring/Drawing Conclusions *(weight: 3)*	Devise one particular scheme.	Devise a general scheme for either of the operations.
Communicating *(weight: 2)*	Give answers only for a particular case without any attempt to verbalize the method used.	Make some attempt to give a more general approach for solving each problem.

NUMBER AND QUANTITY 69

The Trouble With Tables   TEACHER'S GUIDE

Grade Level: Elementary

Description:

Students are asked to demonstrate their ability to reason in the context of simple arithmetic calculation.

Mathematics:

Math Objects

- [X] Number/Quantity
- [ ] Shape/Space
- [ ] Function/Pattern
- [ ] Chance/Data
- [ ] Arrangement

Math Actions (possible weights: 0 through 4)

- [0] Modeling/Formulating
- [4] Transforming/Manipulating
- [2] Drawing Conclusions
- [0] Communicating

Assumed Mathematical Background:

Students should have some experience with solving problems using arithmetic and logical reasoning.

Core Elements of Performance:

- Fill in a table with arithmetic accuracy while obeying given constraints.
- Be able to handle non-unique and unique values in completing the table.

Using This Task:

Read through the prompt with your students to ensure that they understand the task. You may need to emphasize that they must find unique numbers for each cell in question 2.

Extension:

The provided extension is challenging, particularly since a fraction is needed to complete one of the columns. Students may also be given the task of making up a new table for a classmate to complete, possibly employing a combination of addition and multiplication, or subtraction and division.

The Trouble With Tables

1. Lisa needs to do her homework using a table of numbers and their sums and differences. Unfortunately, she got caught in the rain, and some of the numbers were washed away. Help Lisa recreate the table.

A	3		16		10	
B	1	5		4		
A + B			23	9		8
A − B		7			3	4

2. The second table has only a few numbers left in it, but George remembers that **all** the numbers in the table were different from each other. Fill the table with any appropriate numbers, making sure that you do not use a number more than once for either **C** or **D**, and that the sums (**C + D**) and differences (**C − D**) are also different numbers.

C	7					
D	6	3				
C + D			30			
C − D				10	16	

Name: _____ **Date:** _____

Extension:

This table has products and quotients instead of sums and differences:

G	3		15		16	
H	1	5		4		
G × H			45	12		8
G / H		2			4	2

Re-create the table with any appropriate numbers. You may use the same number twice.

THE TROUBLE WITH TABLES   SOLUTION AND RUBRIC

1.

A	3	12	16	5	10	6
B	1	5	7	4	7	2
A + B	4	17	23	9	17	8
A − B	2	7	9	1	3	4

All but the last column are easily found. In the last column, assuming positive number values, **A** would have to be between 4 and 8, and **B** would have to be a number less than 4, as it must fit between the sum and difference. A small amount of experimentation shows that **A = 6**, **B = 2** will work.

2. The second table is more complicated. At first any appropriate numbers may be chosen. These, however, must be adjusted so that no two of them are the same. A possible solution is

C	7	5	21	14	33	34
D	6	3	9	4	17	15
C + D	13	8	30	18	50	49
C − D	1	2	12	10	16	19

There are many other ways to fill the cells with distinct numbers; the task may be simplified significantly if the numbers in successive columns are selected progressively larger, taking advantage of place value. An alternative reconstruction would produce:

NUMBER AND QUANTITY

C	7	20	19	31	33	42
D	6	3	11	21	17	40
C + D	13	23	30	52	50	82
C – D	1	17	8	10	16	2

G	3	10	15	3	16	4
H	1	5	3	4	4	2
G × H	3	50	45	12	64	8
G / H	3	2	5	3/4	4	2

Extension:

2. In this table, all but the last columns are easily determined. Note that a fraction is needed to complete the fourth column. In the last column, the most reasonable guesses for G are the integer divisors of 8, namely 1, 2, 4, and 8. Checking these possibilities, it is seen that **G = 4, H = 2** works.

	Partial Competency	**Full Competency**
Modeling/Formulating *(weight: 0)*		
Transforming/ Manipulating *(weight: 4)*	Fill in most of the first table **and** some of the second table or Fill all of the first table completely and accurately.	Fill in both tables with accurate results (the numbers need not be distinct in the second table for full level in this category). If question **3** is given as an extension, then correct results (including the fractional answer) should count for extra credit.
Inferring/Drawing Conclusions *(weight: 2)*	Correctly handle instances where the columns are not uniquely determined or Develop an appropriate strategy to ensure that all table entries are different in question **2**.	Satisfy **both** the criteria of being able to handle non-unique values and using different entries for each cell in question **2**.
Communicating *(weight: 0)*		

Broken Measures

TEACHER'S GUIDE

Grade Level: Elementary

Description:

Students measure time, length, and weight to demonstrate their understanding of the arbitrariness of the starting point of any measuring scale.

Mathematics:

Math Objects

- [X] Number/Quantity
- [ ] Shape/Space
- [ ] Function/Pattern
- [ ] Chance/Data
- [ ] Arrangement

Math Actions (possible weights: 0 through 4)

- [0] Modeling/Formulating
- [1] Transforming/Manipulating
- [2] Drawing Conclusions
- [1] Communicating

Assumed Mathematical Background:

Students need a basic understanding of methods used to measure weight, length, and time.

Core Elements of Performance:

- Correctly calculate weight, time, and length using the concept of "difference."
- Devise a general scheme for measuring that is independent of the starting point of the measurement scale.

Using This Task:

Read through the prompt with your students to ensure that they understand the task. Try to keep the classroom discussion during the pre-activity directed toward the idea of difference in weight. Similar questions concerning time and length may be pursued as a part of the pre-activity, but if this is done, then the provided questions should be assigned without further explanation. Many students have difficulty with the required subtraction in question 1a (201–185). This is a key diagnostic issue, and any student who is unsuccessful should be flagged for remedial instruction.

Extensions:

Ask students to describe other situations where the starting point of the measurement scale is not zero.

Name: _____ Date: _____

BROKEN MEASURES

Pre-Activity:

A veterinarian who cares for sick cats and dogs has to be able to weigh them.

As you probably know, it is hard to get a cat or dog to stand still on a scale.

How do you think the "vet" weighs these animals?

Task:

1. **a.** A 6 foot 2 inch tall veterinarian who weighs 185 pounds gets on a scale holding a large cat. The scale reads 201 pounds. How much does the cat weigh?

   b. Patrick owns the cat whose weight you found in question **1a**. Patrick is 4 feet 3 inches tall and weighs 70 pounds. If he were to get on his scale at home with the cat, what would his scale read?

Name: _____ **Date:** _____

2. The clock in your classroom is always seven minutes fast. If the clock reads half past eleven, what is the real time?

3. **Your measuring tape was cut.** Here is a picture of the piece you have left. Figure out a way to measure length with this remaining piece of measuring tape, and explain how you would do it.

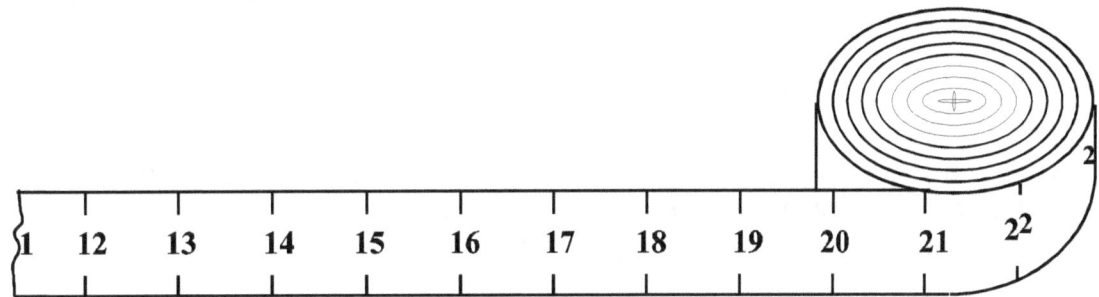

Extension:

Your teacher will measure the height of one person in your class and tell you the height in inches. Using just this information, without rulers or any other measuring instrument, explain how everyone else in the class could estimate his or her own height.

BROKEN MEASURES — SOLUTION AND RUBRIC

Pre-Activity:

The most common approach is for the vet to get on the scale and weigh himself, then to get back on the scale holding the animal and subtract his weight to get the weight of the animal alone.

The classroom discussion should be directed toward the idea of "difference" rather than any mechanical method for restraining the animal!

Task:

1. a. Again, the key element is the idea of difference, that is the scale reading minus the vet's weight, which is 16 pounds.

    b. In this situation, the student must realize that addition is necessary to get a scale reading of 86 pounds.

2. The correct time is 11:23.

3. If you try to use this piece of ruler to measure something that you know is one inch long and you line up the left end of the broken ruler with one end of the object, the other end of the object will be at the 12 mark on the broken ruler. We can use the broken measure if we remember to subtract 11 inches from what we read on it.

Extension:

Each student can stand next to the measured student and estimate how many inches taller or shorter they are.

	Partial Competency	Full Competency
Modeling/ Formulating *(weight: 0)*		
Transforming/ Manipulating *(weight: 1)*	Correctly calculate the values in either **1** or **2**.	Correctly calculate the values in both **1** and **2**.
Inferring/Drawing Conclusions *(weight: 2)*	Devise a mechanical scheme in **3** (e.g., re-number the ruler).	Devise a more general scheme for re-zeroing the ruler in **3**, which may use the concept of "difference."
Communicating *(weight: 1)*	Give a fragile explanation of the measurement schemes.	Explain the measurement schemes clearly and expansively, particularly including a discussion of the use of "difference."

NUMBER AND QUANTITY 79

COUNTING OFF

TEACHER'S GUIDE

Grade Level: Elementary

Description:

Students demonstrate their ability to design counting schemes and to recognize the upper limits of a particular counting situation.

Mathematics:

Math Objects

[X] Number/Quantity [ ] Shape/Space [ ] Function/Pattern

[ ] Chance/Data [ ] Arrangement

Math Actions (possible weights: 0 through 4)

[2] Modeling/Formulating [1] Transforming/Manipulating

[3] Drawing Conclusions [3] Communicating

Assumed Mathematical Background:

Students must be able to count by twos, by threes, and by fives. They also need experience in problem solving using logical reasoning.

Core Elements of Performance:

- Design counting schemes using the given directions.
- Recognize the upper limits of a particular counting situation.

Using This Task:

Read through the prompt with your students to ensure that they understand the task. Make sure that students are clear on the meaning of "less than 35." They should also be reminded that each question is based on the same group of children, and each answer must reflect the results of the previous answers.

Extensions:

The required verbalization of the result in the last question may be viewed as an extension for those students with limited writing skills. It may also be used as a culminating homework assignment after the numerical parts of the task have been addressed.

Name: _____   Date: _____

COUNTING OFF

A group of children stood in a line. There were *less than 35* children in the group.

1. The children counted off by twos. The first child said 1, the next said 2, the next said 1, and so on. The last child said 1.

   List the different numbers of children that could be in the group.

2. Then **the same group** of children counted off by threes: 1, 2, 3, 1, 2, 3.... The last child said 1.

   Now what are the different numbers of children that there could be in the group?

Name: _____ **Date:** _____

3. Finally **the same group** of children counted off by fives. This time the last child said 3.

   What is the possible number of children in the group?

4. Use your answers from questions **1**, **2**, and **3** to determine how many children are in this group.

   Write a letter to a friend who is absent from school explaining how you went about finding the exact number of children.

COUNTING OFF — SOLUTION AND RUBRIC

1. The number of children could be any odd number less than 35.

2. The number of children must be one more than a multiple of three (1, 4, 7, 10, 13,...,34). When this finding is combined with the previous result, the only possible numbers are 1, 7, 13, 19, 25, and 31.

3. This tells us that the number of children must end in a 3 or an 8. When this finding is combined with the results of **1** and **2**, it eliminates all the numbers ending in 8 (answer must be odd) and all the numbers less than 35 ending in 3 except 13 (one more than a multiple of 3).

4. Since we are dealing with the same group of children in each question, the number of children in the group must appear in all the listed sets of possibilities; the only number that meets this requirement is 13.

	Partial Competency	Full Competency
Modeling/ Formulating *(weight: 2)*	Design a correct, organized counting scheme for some, but not all of the questions	Design a correct, organized counting scheme for each question.
Transforming/ Manipulating *(weight: 1)*	Calculate some responses correctly.	Calculate all responses correctly.
Inferring/Drawing Conclusions *(weight: 3)*	Reach correct conclusions and recognize the upper limit while designing the counting scheme for each individual question.	Combine the results from **3** to deduce the correct number of children in **4**.
Communicating *(weight: 3)*	Write a letter that conveys the correct numerical answers for **1**, **2**, and **3**.	Write a letter that provides some explanation of the process leading to the final number of children in **4**.

Network News

TEACHER'S GUIDE

Grade Level: Elementary

Description:

Students demonstrate, through arithmetic computation, the relation between addition and its inverse, multiplication and its inverse, and the effects of the distributive law.

Mathematics:

Math Objects

- [X] Number/Quantity
- [ ] Shape/Space
- [ ] Function/Pattern
- [ ] Chance/Data
- [ ] Arrangement

Math Actions (possible weights: 0 through 4)

- [2] Modeling/Formulating
- [2] Transforming/Manipulating
- [3] Drawing Conclusions
- [2] Communicating

Assumed Mathematical Background:

This task assumes basic arithmetic instruction in addition, subtraction, multiplication, and division. It also assumes an understanding of inverse operations, the effect of order of operations with integers, and the distributive property.

Core Elements of Performance:

- Arrive at a correct result for each branch of the network.
- Design a new path for the network that meets the criteria.
- Exhibit an understanding of the arithmetic generalization described by the network, the inverse relationship of the arithmetic processes involved, and the effect of the distributive rule on the results.

Using This Task:

This task should be used with upper elementary students who have already done *Multiplication Rings*, as they need the experience of describing the working of the rings before they tackle the more complicated network. There may be a need to do some review of inverse operations, as often these processes are taught in relative isolation, and students do not have a clear concept of how one process undoes the other.

Extension:

A truly challenging extension is to design a branch that uses all four arithmetic processes and that works in either direction, that is one in which it does not matter whether you start at the **IN** box or the **OUT** box.

Name: _____ Date: _____

NETWORK NEWS

Here is a special kind of network—any number that goes into the **IN** box gets changed around in complicated ways. Finally the number that comes out the **OUT** box is the same as the number that went in.

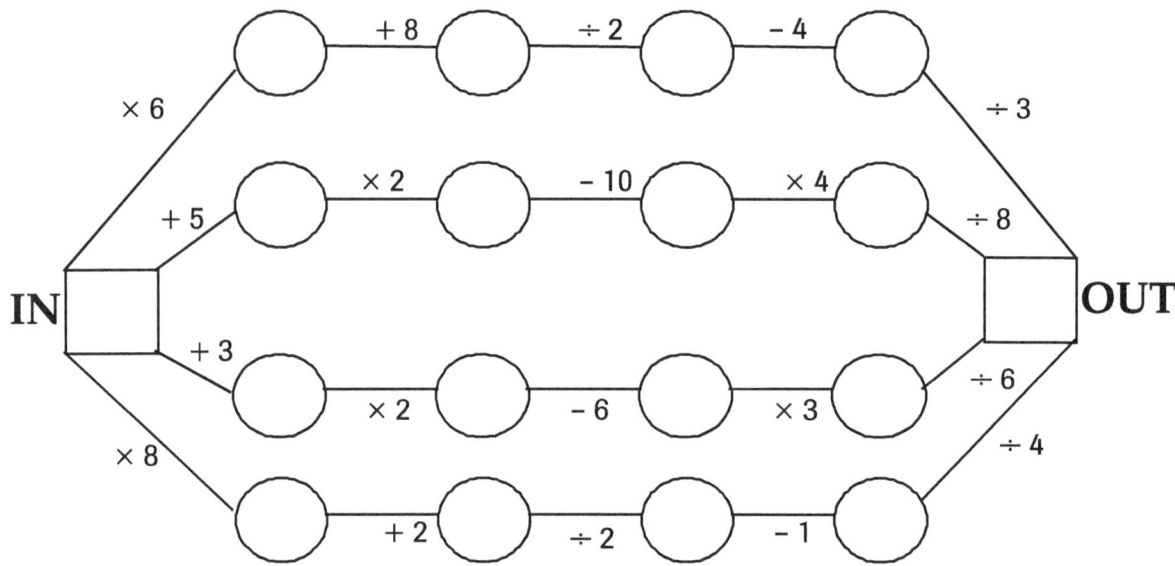

1. Try this network, putting several different numbers into the **IN** box and following along each of the four different paths from **IN** to **OUT**. Do you always get out the same number you put in?

Name: _____ **Date:** _____

2. Add another path to this network that starts at the **IN** box and goes to the **OUT** box without changing the original number.

3. Explain how you think this network works.

Network News

SOLUTION AND RUBRIC

The intent of this task is to have students demonstrate arithmetic computation skills, specifically the relation between addition and its inverse, multiplication and its inverse, and the effects of the distributive law.

It takes some care to follow the operations in the network correctly, but in the end, the "in" number should be identical to the "out" number for any branch of the network.

Students will often manage to design a workable path by using a multiple of the numbers in any given branch, without changing the order of the arithmetic processes. Another approach is to design a branch that uses only addition and subtraction, or multiplication and division. It may take some work to balance the terms, but the result is less interesting than the combined operations demonstrated in the given branches.

Why do these networks work? To look at the problem algebraically, each branch of the network has either this structure

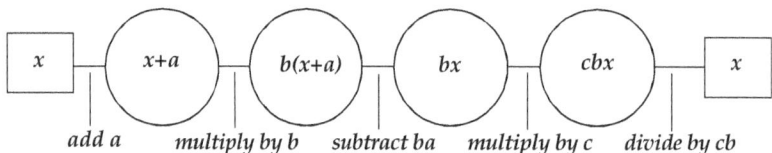

or this structure

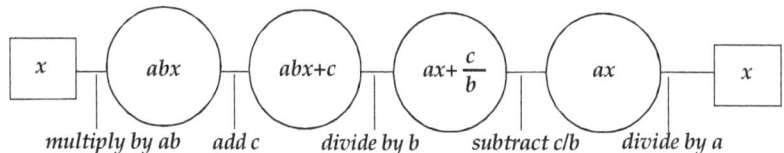

If you use any numbers *a*, *b*, and *c*, provided none of them is zero, this procedure guarantees that the number *x* that emerges is simply the number that was put in.

Elementary students may describe this as a process in which the multiplications and divisions "balance out" or "cancel out." Even though the numbers that are being added and subtracted are not the same, whichever is done first is immediately acted upon by a multiplication or a division, which has the net effect of enlarging its value or reducing it so that it matches the net result of its inverse operation.

	Partial Competency	Full Competency
Modeling/ Formulating *(weight: 2)*	Make a new branch with **only** addition/subtraction **or** multiplication/division.	Make a new branch with **both** addition/subtraction **and** multiplication/division.
Transforming/ Manipulating *(weight: 2)*	Make no more than one error per branch.	Do all computations correctly.
Inferring/Drawing Conclusions *(weight: 3)*	Argue that opposite operations compensate for each other.	Additionally, take the distributive property into account.
Communicating *(weight: 2)*	Give a somewhat complete verbal, iconic, or symbolic explanation.	Give a clear, complete explanation, using more than one representation.

Piece of String

Teacher's Guide

Grade Level: Elementary

Description:

This task assesses student ability to measure length using non-standard units and to make reasonable estimates of the linear dimensions of the various body parts.

Mathematics:

Math Objects

[X] Number/Quantity   [ ] Shape/Space   [ ] Function/Pattern

[ ] Chance/Data   [ ] Arrangement

Math Actions (possible weights: 0 through 4)

[0] Modeling/Formulating   [2] Transforming/Manipulating

[3] Drawing Conclusions   [2] Communicating

Assumed Mathematical Background:

Students should have experience in measuring length and also in making estimates.

Core Elements of Performance:

- Correctly order measurements in a table.
- Make reasonable estimates of the linear dimensions of various body parts.
- Make appropriate modifications to original estimates based on new information.

Using This Task:

Read through the prompt with your students to ensure that they understand the task. It may be useful to highlight that the answers for each part of question 2 are dependent on the string being the same length as the height of the person in the picture. Students may also need to be reminded to fill in both the "more than" and the "less than" columns in question 1b, if the string does not go around an exact number of times.

Extension:

For students with limited writing skills, the written explanations required in question 2 will be an extension.

Students can also be asked to describe other situations where the ratios between the linear dimensions of various body parts are not usual.

Materials Needed:

String, scissors

Name: _____   Date: _____

PIECE OF STRING

1. Cut a piece of string that is as long as the distance from the tip of your toes to the top of your head.

    a. How many times do you think this string would go around your head? around your waist? around your wrist? around your ankle? around your pinky?

    Make an estimate and put your results into the table below:

How many times around your	Estimate
head	
waist	
wrist	
ankle	
pinky	

Name: _____   Date: _____

b. Now use your piece of string to measure each one, and fill in the following table. If the string goes around a whole number of times, use the "exactly" column. If the string does **not** go around a whole number of times, fill in the "more than ___ times" and "less than ___ times" columns.

How many times around your	Measurement	
	exactly	more than BUT less than
head		
waist		
wrist		
ankle		
pinky		

c. How well do your estimates compare with your real measurements?

Name: _____   Date: _____

2. What do you think the results would be if the people represented by the following pictures did the same experiment? (Remember that in each case the string will be the same length as the height of the person in the picture.) Put your answers into the table and explain them.

How many times around the	exactly
head	
waist	
wrist	
ankle	
pinky	

Explanation:

Name: _____   Date: _____

How many times around the	exactly
head	
waist	
wrist	
ankle	
pinky	

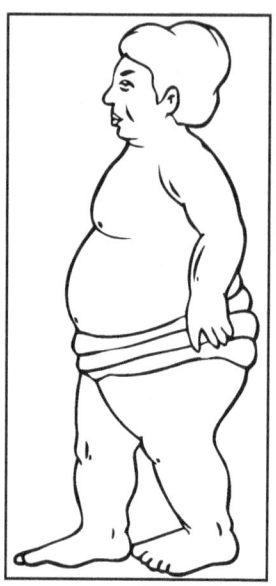

Explanation:

How many times around the	exactly
head	
waist	
wrist	
ankle	
pinky	

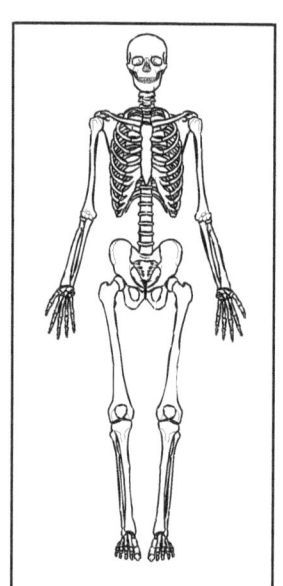

Explanation:

A Piece of String

SOLUTION AND RUBRIC

1. The results of fourth grade student's measurements will be close to the following:
   - head: about 3 times
   - waist: about 3 times
   - wrist: about 10 times
   - ankle: about 7 times
   - pinky: about 30 times

   Their estimates may be substantially different.

2. Students are expected to recognize the differences in proportion between their own bodies and those of a baby, a sumo wrestler, and a skeleton. Their answers should reflect observations such as
   - a baby's head is in larger proportion to the rest of the body, thus the head ratio estimate for the baby should be smaller than the fourth grader;
   - the sumo wrestler has a very large stomach, so his waist ratio should be smaller than the fourth grader;
   - the skeleton has a significantly smaller waist than a complete body, so this estimate should be larger than the fourth grader.

	Partial Competency	Full Competency
Modeling/ Formulating *(weight: 0)*		
Transforming/ Manipulating *(weight: 2)*	Show partial success in completing the table for **1**.	Put all measurements correctly into the table for **1**, and show evidence of understanding the relative positions of the "exactly" integers and the "more than/less than" integers (e.g., no fractional answers are given).
Inferring/Drawing Conclusions *(weight: 3)*	Make estimates that run in approximately the correct order (smallest estimates for the head and waist, largest for the pinky).	Make appropriate modifications to the original estimates that appropriately reflect the body characteristics of each situation in **2**.
Communicating *(weight: 2)*	Communicate answers in **1** that are not consistent as to "exactly" versus "more than/less than" or Provide answers for **2** but give no explanation of assumptions.	Communicate consistent answers in **1** and give clear explanations for the answers in **2**, especially where they differ from the results for the student's own body

3

Shape and Space

Students come to school with a basic sense of shape, and one of the first formal mathematics activities that occurs is the describing and categorizing of these mathematical objects.

During the primary grades, students should begin to demonstrate

- competence in distinguishing and naming a variety of two-dimensional shapes, and ability to follow directives such as *draw two different kinds of closed figures that have three straight lines;*
- an understanding of the symmetries of shapes, and ability to follow directives such as *find all the lines along which you can fold a paper square so that the two parts lie exactly on top of each other.*

During the elementary years we expect students to grow in their ability to demonstrate

- competence in distinguishing and naming additional two dimensional and also three-dimensional shapes, and ability to follow directives such as *draw three different kinds of closed figures that have four straight lines;*
- enhanced understanding of the symmetries of these shapes, and ability to follow directives such as *find all the lines along which you can fold a paper hexagon so that the two parts lie exactly on tops of each other.*

As students transition into the upper elementary grades, we expect that they will be able to work with increased facility in performing such calculations as finding the area, perimeter, and volume of more complex geometric shapes, and will also appreciate the relationship between geometry and algebra as evidenced in investigations of coordinate geometry.

The tasks in this chapter provide the opportunity to demonstrate these abilities in a variety of ways. Some of the tasks are a blending of geometry and pattern, or geometry and arrangement. Some require more computation than others; some have a heavy demand on modeling and inference. Others require that the student communicate their mathematical understanding in various ways. By choosing tasks that are weighted more heavily in any of these four process areas, you will be better able to individualize the assessment, and respond to the varying competencies and weaknesses of each student.

In most cases it will be necessary to scribe the answers for kindergarten and early Grade 1 students, but every effort should be made to have the students draw their own diagrams and move to expressing their own understanding in writing as soon as possible.

Also, it is important that teachers be aware of the variance in ability of students to transition between two-dimensional and three-dimensional space. This is a skill that is often not solidly grounded until middle school or later; it is dependent on the functioning of the prefrontal cortex, which typically is not fully developed until early adolescence.

Grassy Parks

Grade Level: Primary

TEACHER'S GUIDE

Description:

This task assesses student ability to distinguish between the perimeter and the area measurements of three plane shapes in a real-world situation.

Mathematics:

Math Objects

- [X] Number/Quantity
- [X] Shape/Space
- [ ] Function/Pattern
- [ ] Chance/Data
- [ ] Arrangement

Math Actions (possible weights: 0 through 4)

- [0] Modeling/Formulating
- [2] Transforming/Manipulating
- [2] Drawing Conclusions
- [2] Communicating

Assumed Mathematical Background:

Student should have some experience in counting blocks on a grid.

Core Elements of Performance:

- Determine the size (area) of the three shapes and decide which is the largest.
- Determine the distance around (perimeter) the three shapes and decide which is the largest.
- Reconcile the fact that two of the shapes have the same area.

Using This Task:

Read through the prompt aloud with your students to ensure that they understand the task, and answer any questions that arise.

If this task is utilized for informal classroom use in kindergarten or during the early part of Grade 1, you may want to do a pre-activity that involves counting squares on a grid. For some very young students, it may also be necessary to remind them that there is a grid square "behind" each of the shaded pieces. If students seem stuck on their visual perception for the size of each shape, you might suggest that they first guess which park is the biggest just by looking at it, and then find out for sure which has the biggest area by counting the number of grassy squares in each shape.

Extension:

Grassy Parks may be extended for either instructional or assessment purposes by asking students to write a prose explanation that describes any relationship they see between perimeter and area. This considerably increases the communication demand of the task.

Grassy Parks

Here are pictures of the shapes of three grassy parks.

Name: _____ **Date:** _____

1. In which park would you have the most space to play?

   Explain why you think so.

2. In which park would you need to take the most steps to walk all around its edge?

   Explain why you think so.

Grassy Parks — SOLUTION AND RUBRIC

1. A complete response will include the fact that both Park B and Park C have the same amount of space, and that they have the most space because they have the most squares, namely 7.

2. Students should identify Park B as having the largest perimeter (16 steps) versus 10 steps for Park A and 14 steps for Park C.

	Partial Competency	Full Competency
Modeling/ Formulating *(weight: 0)*		
Transforming/ Manipulating *(weight: 2)*	Student correctly determines the size of one or two of the shapes. Student correctly determines the distance around one or two of the shapes.	Student correctly determines the size of all three shapes. Student correctly determines the distance around all three shapes.
Inferring/Drawing Conclusions *(weight: 2)*	Student finds **either** the shape with the largest area **or** the shape with the longest perimeter.	Student determines both the largest area and the largest perimeter, and is able to reconcile the fact that two of the shapes have the same area.
Communicating *(weight: 2)*	Student gives a partial explanation of **either** the park with the most space **or** the park with the largest perimeter, using limited vocabulary.	Student gives a complete, clear prose justification for the choice of the park with the most space and the park with the largest perimeter. Some students may make explicit comparison with the size of the other parks.

SHAPE AND SPACE

STICKERS

TEACHER'S GUIDE

Grade Level: Primary

Description:

This task requires students to recognize shape and color patterns in a row of "stickers." Students are asked to draw shapes to complete the pattern and to describe in words the patterns they see.

Mathematics:

Math Objects

| ☐ Number/Quantity | ☒ Shape/Space | ☒ Function/Pattern |
| ☐ Chance/Data | ☐ Arrangement | |

Math Actions (possible weights: 0 through 4)

| [0] Modeling/Formulating | [1] Transforming/Manipulating |
| [2] Drawing Conclusions | [2] Communicating |

Assumed Mathematical Background:

This task requires experience with basic pattern recognition.

Core Elements of Performance:

- Complete a given geometric pattern.
- Identify and describe multiple patterns.

Using This Task:

Read through the prompt aloud with your students to ensure that they understand the task, and answer any questions that arise.

In an informal classroom situation, you may wish to remind students to fully describe their patterns, using appropriate geometric vocabulary.

If Kindergarten and early Grade 1 students are having difficulty drawing the objects, you can photocopy the pattern strips onto adhesive paper and cut them up into individual stickers. It may also be helpful to model this task with attribute blocks or other appropriate manipulatives.

Name: _____ Date: _____

Stickers

A roll of stickers has stickers of two different colors and five different shapes. The stickers are in a repeating pattern, but three of the stickers have been peeled off from the roll.

1. Draw in the missing stickers.

2. Describe a pattern you see in the stickers.

3. Describe another pattern you see in the stickers.

STICKERS

SOLUTION AND RUBRIC

The two patterns that appear in the roll of stickers are

- the color of the objects alternates between black and white;
- the shape of the objects repeats in a 5-shape cycle: square, circle, star, triangle, oval.

The three blank spaces should be filled with a white star, a black square, and a white triangle.

	Partial Competency	Full Competency
Modeling/ Formulating *(weight: 0)*		
Transforming/ Manipulating *(weight: 1)*	Some but not all of the stickers are correctly drawn.	All of the stickers are correctly drawn. or The stickers drawn are incorrect but entirely consistent with the student's verbal description of a pattern.
Inferring/Drawing Conclusions *(weight: 2)*	The student shows evidence of having discovered only one of the two patterns (color or shape).	The student shows evidence of having discovered both of the patterns (color and shape).
Communicating *(weight: 2)*	The student's verbal descriptions of the patterns are somewhat understandable, but not well-written.	The student's verbal descriptions of the patterns are well-written.

Shirts in the Mirror

TEACHER'S GUIDE

Grade Level: Primary

Description:

Students are asked to explore the letters that look the same and the letters that look different when seen as a reflection in a mirror. The task progresses from single letters to whole words.

Mathematics:

Math Objects

- [ ] Number/Quantity
- [x] Shape/Space
- [ ] Function/Pattern
- [ ] Chance/Data
- [ ] Arrangement

Math Actions (possible weights: 0 through 4)

- [2] Modeling/Formulating
- [1] Transforming/Manipulating
- [2] Drawing Conclusions
- [1] Communicating

Assumed Mathematical Background:

This task assumes basic understanding of symmetry and reflection.

Core Elements of Performance:

- Determine which alphabet letters have reflection symmetry.
- Draw the reflected image of groups of letters.

Using This Task:

Read through the prompt aloud with your students to ensure that they understand the task, and answer any questions that arise.

Ideally, students should not use actual mirrors while working this task, as it is designed to promote visualization and a more conceptual sense of symmetry. However, if students seem "stuck" as to how to get into the problem, it may be helpful to demonstrate one or two examples using a mirror.

Kindergarten students should be able to do questions 1 and 2; older students will enjoy the challenge of questions 4 and 5.

Extension:

Older students should be able to come up with other words that have reflection symmetry.

Name: _____ Date: _____

SHIRTS IN THE MIRROR

Have you ever noticed that when you look in a mirror, the right side of your face is on the left, and the left side of your face is on the right?

So what happens if you look in the mirror while wearing a T-shirt that has a letter on it?

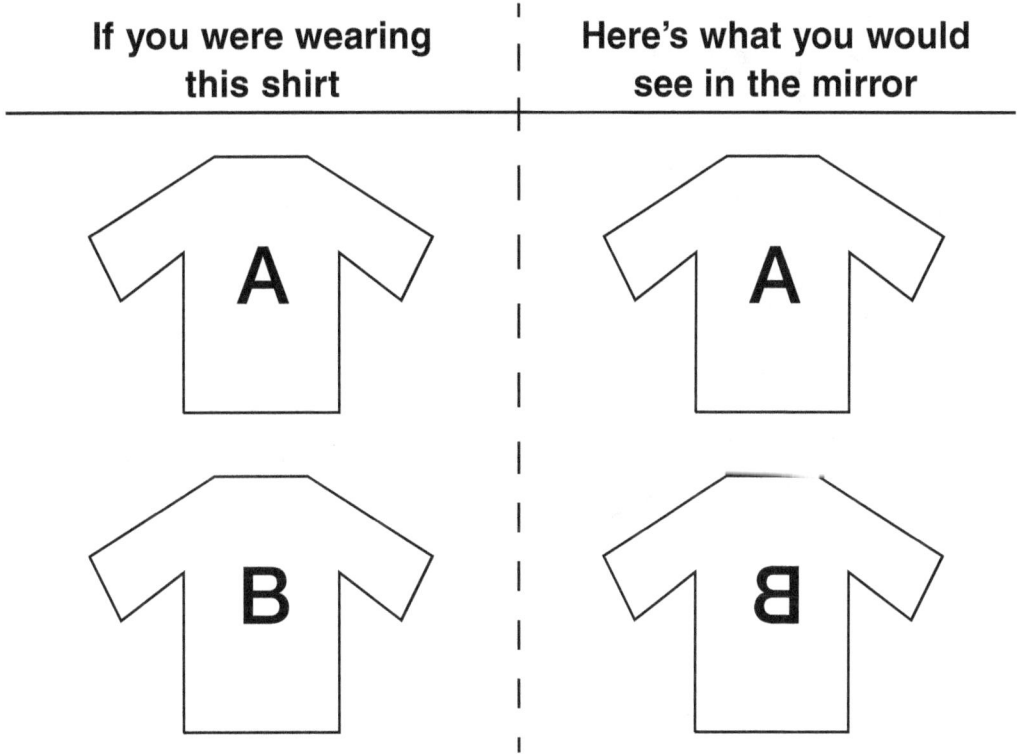

Notice that letter **A** looks **the same as itself** in the mirror, but letter **B** looks **different** in the mirror.

Name: _____ Date: _____

Here are the rest of the letters of the alphabet.

C	D	E	F	G	H	I	J	K	L	M	N
O	P	Q	R	S	T	U	V	W	X	Y	Z

1. Fill in each of these shirts with a letter that would look the same as itself in the mirror. Use a different letter for each shirt.

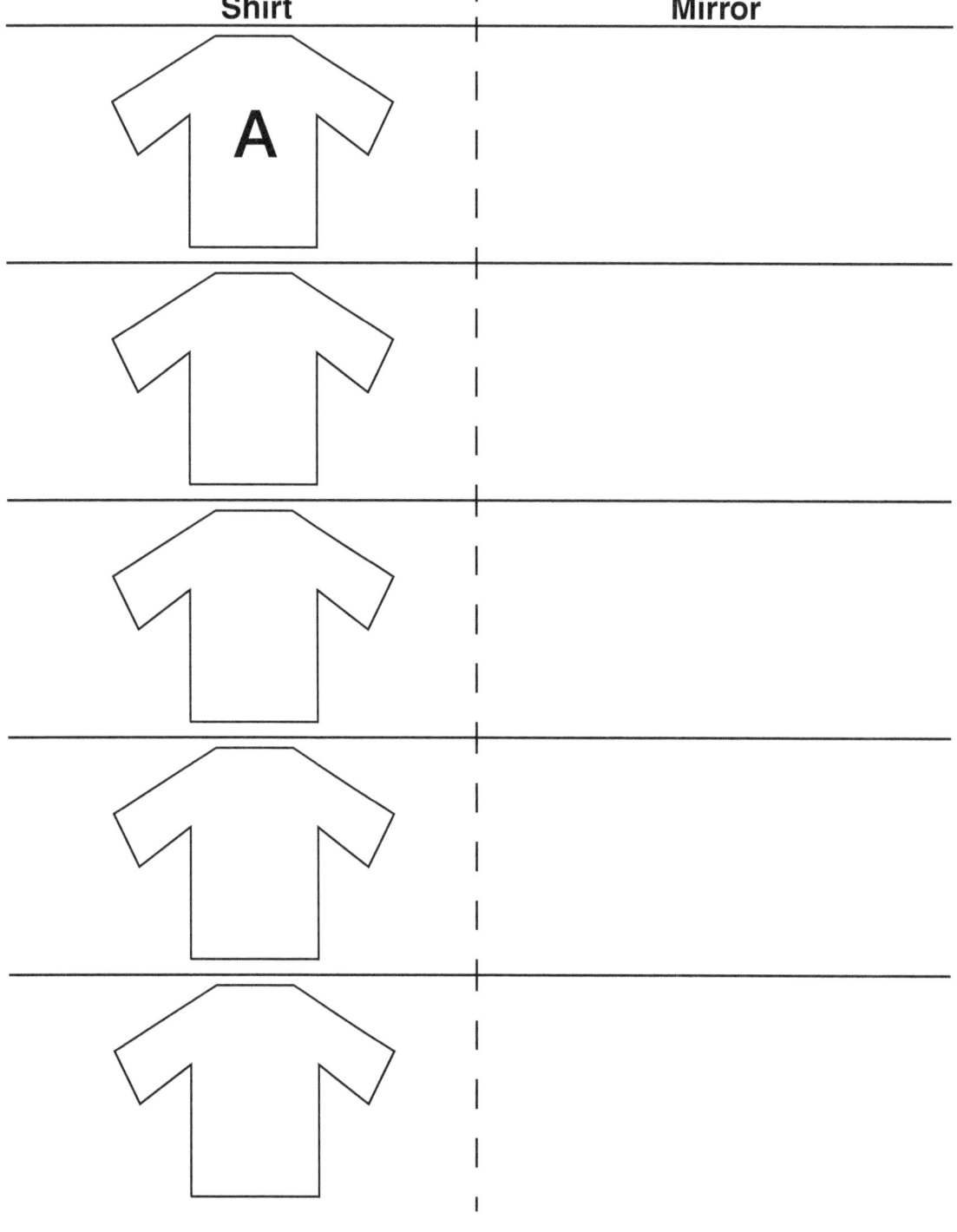

Name: _____ Date: _____

2. Fill in these shirts with letters that look **different** in the mirror. Also, draw what the letters look like in the mirror. Use a different letter for each shirt.

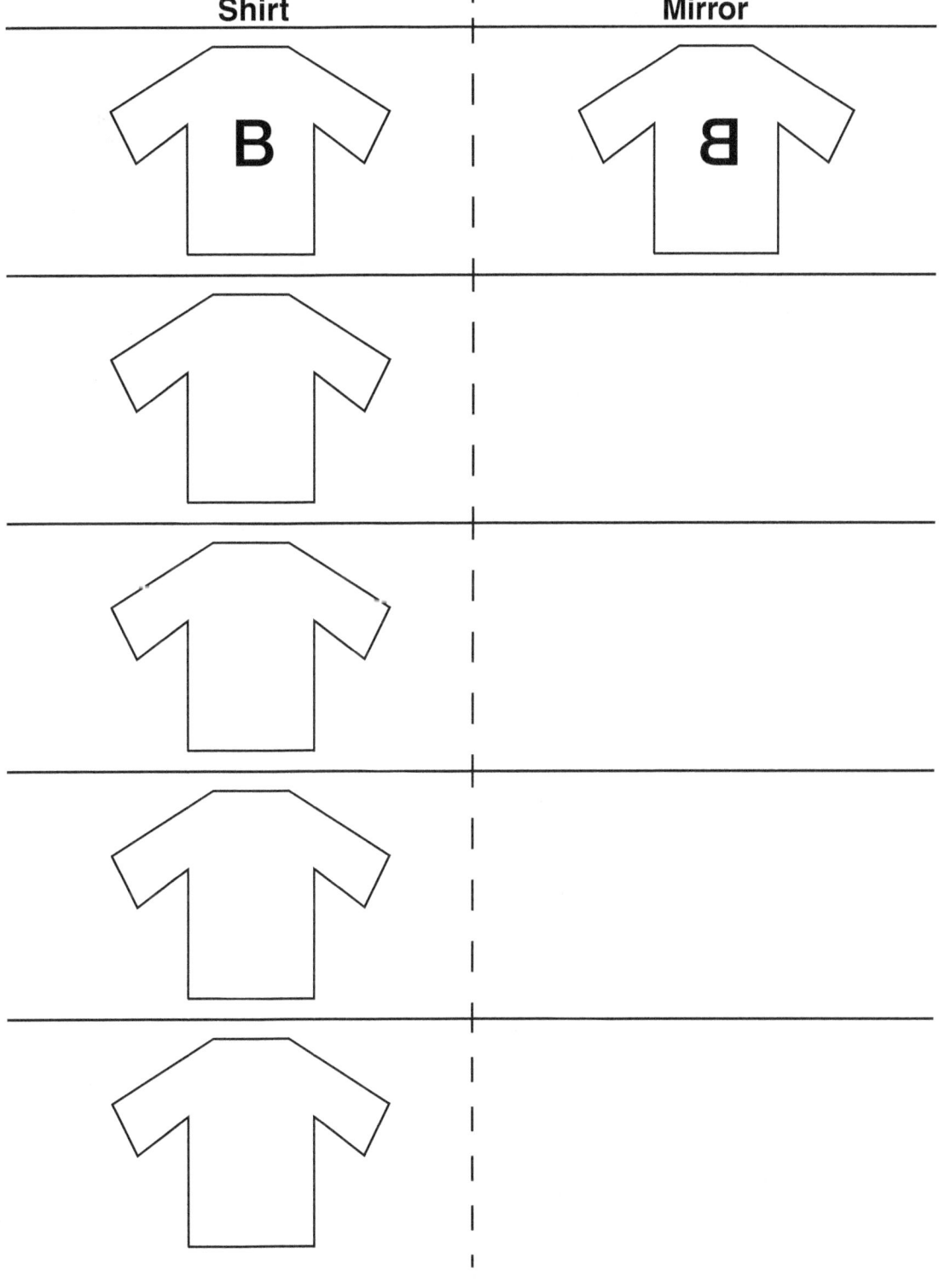

Name: _____ Date: _____

3. Here are pictures of some T-shirts from colleges. Draw what each would look like in the mirror.

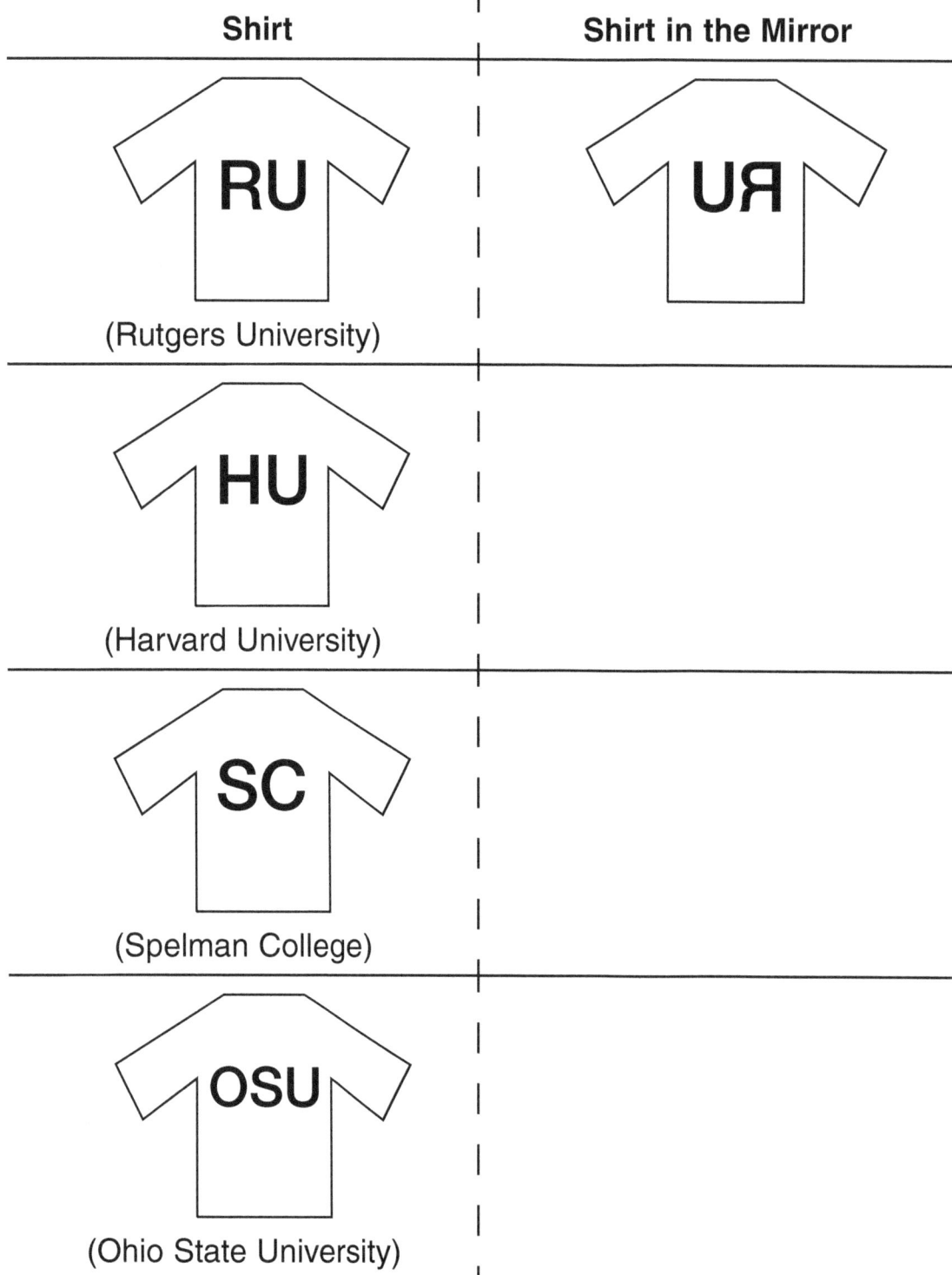

Name: _____ Date: _____

4. Suppose you were wearing a T-shirt containing the word WOW. What would you see when you looked at your shirt in the mirror?

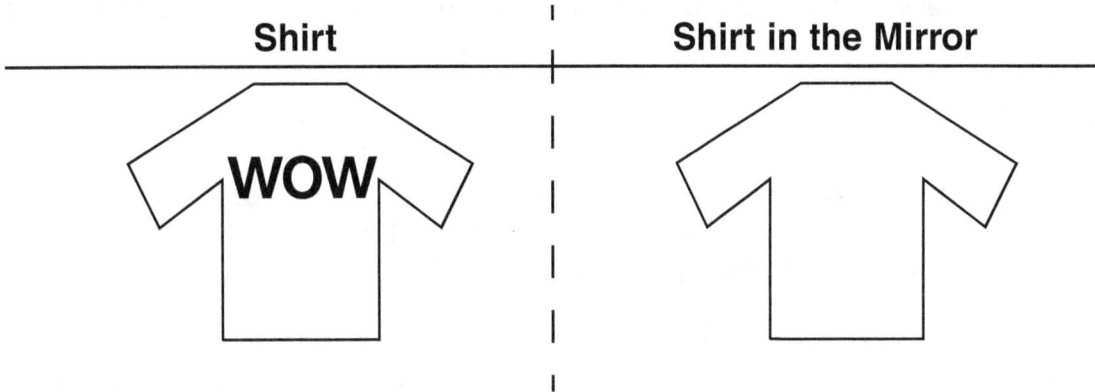

Extension:

5. Can you think of any other whole words or names that look the same in the mirror?

Shirts in the Mirror
Solution and Rubric

1. Any of these letters could be used: **H, I, M, O, T, U, V, W, X, Y**.
2. Any of these letters could be used: **C, D, E, F, G, J, K, L, N, P, Q, R, S, Z**.
3. Here's what the letters on each shirt would look like in the mirror: **UH, ЭS, USO**.
4. The letters would look the same in the mirror: **WOW**.

Extension:

5. Any word that is formed from the letters **A, H, I, M, O, T, U, V, W, X**, and **Y**, and that is spelled the same forward and backward will look the same in the mirror. Some examples include **MOM, TOT, TOOT,** and **OTTO**.

	Partial Competency	**Full Competency**
Modeling/ Formulating *(weight: 2)*	The student devises procedures for making the mirror images that are only partially correct (for example, handling single letters correctly but groups of letters incorrectly).	The student devises procedures for making the mirror images that are fully correct.
Transforming/ Manipulating *(weight: 1)*	Some errors are made in the formation of mirror images.	Student consistently displays the ability to form correct mirror images.
Inferring/Drawing Conclusions *(weight: 2)*	For questions **2**, **3**, and **5**, the student's choices of letters and words are not fully correct (for example, the student makes some incorrect choices of letters or leaves some of the shirts blank).	For problems **2**, **3**, and **5**, the student correctly chooses letters and words that meet the requirements of the problem.
Communicating *(weight: 1)*	Letters and mirror images are drawn in such a way that the student's intended answer is sometimes not clear.	Letters and mirror images are drawn in such a way that the student's intended answer is always clear (whether correct or not).

SHAPE AND SPACE

Does It Fit?

TEACHER'S GUIDE

Grade Level: Elementary

Description:

This task assesses student understanding of rotation and reflection symmetry.

Mathematics:

Math Objects

- [ ] Number/Quantity
- [X] Shape/Space
- [ ] Function/Pattern
- [ ] Chance/Data
- [X] Arrangement

Math Actions (possible weights: 0 through 4)

- [2] Modeling/Formulating
- [0] Transforming/Manipulating
- [2] Drawing Conclusions
- [2] Communicating

Assumed Mathematical Background:

This task assumes an understanding of the attributes of regular plane shapes, including their symmetry.

Core Elements of Performance:

- Devise consistent and appropriate counting schemes.
- Explicitly state counting assumptions or methods.

Using This Task:

Read through the prompt aloud with your students. Once the prompt has been read, this task should be presented with no additional instruction. If students are having extreme difficulty, allow them to cut out the signs and manually rotate them.

Extension:

Talk about others signs that have different polygonal shapes (e.g., hexagons and octagons) and calculate how many ways they can fit into a similarly-shaped envelope. Older elementary students may be able to generalize to polygons with *n* number of sides.

Name: _____ Date: _____

Does It Fit?

A factory makes two kinds of signs like the ones shown below for the highway department:

1. Each "Danger" sign is shipped in a triangular padded envelope into which it just fits. How many different ways can a sign be put into its envelope? Explain your answer.

2. Each "School Zone" sign is shipped in a square padded envelope into which it just fits. How many different ways can a sign be put into its envelope? Explain your answer.

Does It Fit?

SOLUTION AND RUBRIC

1. The main issue is how the positions will be counted. This raises two questions.

    Does putting the sign in the envelope as pictured, or rotated, count as more than one position? Since usually only one side of the envelope is unsealed and has a flap on it, rotating the sign should constitute different positions.

    Furthermore, can the sign be inserted in the envelope face down to get more choices? Students should recognize the possibility that the sign could be placed "upside down" in the envelope. If "right-side-up" and "upside-down" are counted as distinct positions, this doubles the number of possibilities.

    The "Danger" sign can go into the envelope face up in three ways, and face down in three ways—six ways in all.

2. The considerations are the same as in question 1. The "School Zone" sign can go into its envelope face up in four ways, and face down in four ways—eight ways in all.

	Partial Competency	**Full Competency**
Modeling/ Formulating *(weight: 2)*	Devise appropriate numerical counting schemes for **1** or **2**.	Devise appropriate numerical counting schemes for **1** and **2**.
Transforming/ Manipulating *(weight: 0)*		
Inferring/Drawing Conclusions *(weight: 2)*	Meet some but not all of the criteria for full level.	Consider the distinction between "right-side-up" and "upside-down." Verify that no position is counted more than once. Use a consistent scheme between **1** and **2**.
Communicating *(weight: 2)*	State answers clearly.	Give a clear prose explanation of the answers. Explicitly state counting assumptions or methods.

Mirror, Mirror — TEACHER'S GUIDE

Grade Level: Elementary

Description:

This task assesses student understanding of rotation and reflection symmetry.

Mathematics:

Math Objects

- [ ] Number/Quantity
- [X] Shape/Space
- [ ] Function/Pattern
- [ ] Chance/Data
- [X] Arrangement

Math Actions (possible weights: 0 through 4)

- [1] Modeling/Formulating
- [2] Transforming/Manipulating
- [2] Drawing Conclusions
- [1] Communicating

Assumed Mathematical Background:

This task assumes a basic understanding of rotation and reflection symmetry.

Core Elements of Performance:

- See two or more symmetries in squares and equilateral triangles.
- Provide clear, correct diagrams of the various mirror positions.
- Recognize that the lengths in the mirror are the same as the lengths in the original drawing and articulate some symmetry argument.

Using This Task:

Read through the prompt with your students to ensure that they understand the task.

If students have not used mirrors before, it will be useful to let them "play" with the mirrors for a while before doing the task.

Advise students to be careful in drawing their squares in order to preserve the symmetry.

Extension:

Have students write a prose explanation for the situation posed in question 3, covering both possibilities. This substantially increases the communication demand of the task.

Materials Needed:

Rulers, small hand mirrors or double-sided reflective strips.

Name: _____   Date: _____

Mirror, Mirror

Draw a square on a piece of paper. Now take a mirror and hold it so that its edge is on the paper.

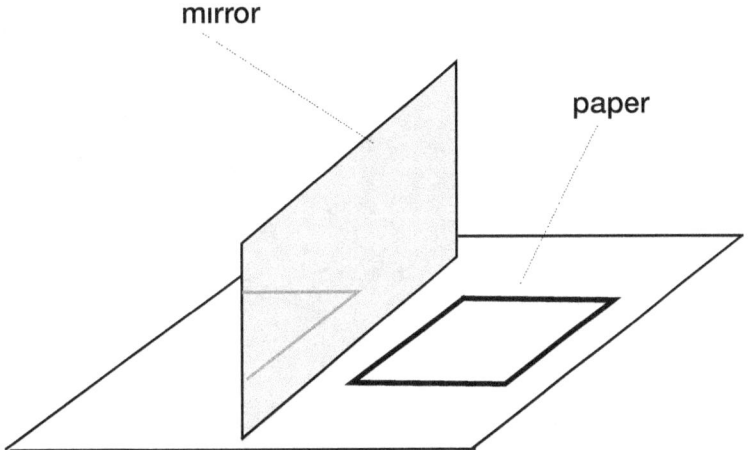

1. Find all the ways of placing the mirror on the paper so that what you see on the paper **together** with what you see in the mirror makes **one** complete square which is exactly the same as the one you drew. (Notice that if you leave the mirror in the same place as is shown in the diagram, you will see **two** squares, not one.)

Name: _____ **Date:** _____

2. Draw a triangle and do the same thing with the mirror. How many different ways can you hold the mirror so that what you see on the paper **together** with what you see in the mirror is the same triangle as you have drawn on the paper?

3. Draw another triangle that lets you place the mirror in **either** more ways or fewer ways.

Mirror, Mirror

SOLUTION AND RUBRIC

Two conditions are necessary for the image in the mirror to be identical to the part that is covered by it: each side or each part of a side covered by the mirror must be equal in length to the corresponding side or part of a side in front of the mirror.

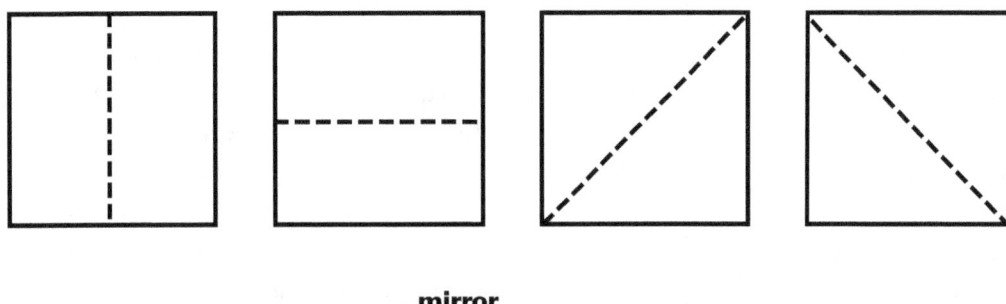

‑ ‑ ‑ ‑ ‑ ‑ ‑ ‑ ‑ mirror

1. Given this condition, there are three possible correct answers:

   a. The mirror goes through the centers of opposite sides or through opposite corners (2 ways).

   b. The mirror is positioned horizontally, vertically, and along the two diagonals (4 ways).

   c. If the mirror is double-faced, in each of the above positions the mirror can face in two directions (8 ways).

2. Given a random scalene triangle, there is **no** position of the mirror that will duplicate the entire triangle. If the chosen triangle is isosceles, there are either one or two positions, depending on whether changing the direction of the mirror face can be considered. If the chosen triangle is equilateral, there are either three or six positions.

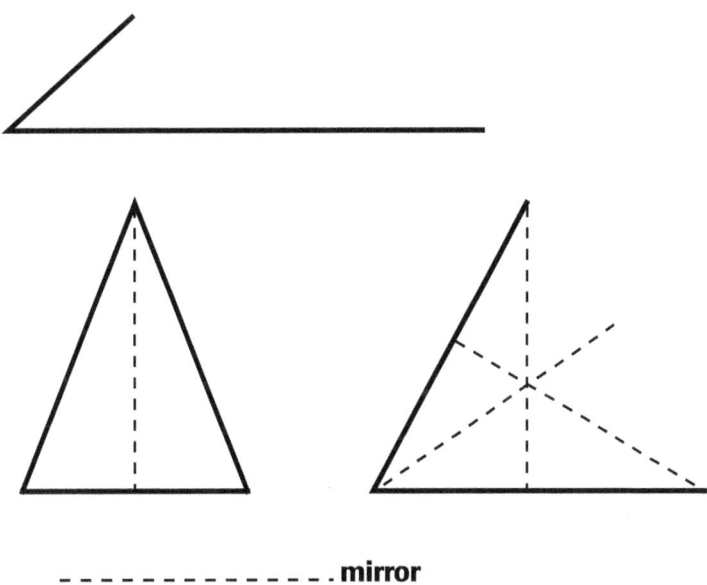

‑ ‑ ‑ ‑ ‑ ‑ ‑ ‑ ‑ ‑ mirror

	Partial Competency	**Full Competency**
Modeling/ Formulating *(weight: 1)*	Devise a model that retains some of the linear measurements.	Additionally, recognize that the lengths in the mirror are the same as the lengths in the original and make some symmetry argument.
Transforming/ Manipulating *(weight: 2)*	Provide a descriptive justification for the chosen position of the mirror.	Additionally, identify at least two different kinds of triangles in **2**.
Inferring/Drawing Conclusions *(weight: 2)*	See one symmetry in the square and/or one symmetry in the triangles (provided a scalene triangle is not drawn).	See two or more symmetries in squares and equilateral triangles.
Communicating *(weight: 1)*	Provide reasonably clear, correct diagrams of the various mirror positions.	Additionally, provide clear prose where necessary and be able to articulate a consistent argument.

Gardens of Delight

TEACHER'S GUIDE

Grade Level: Elementary

Description:

This task assesses student ability to distinguish the attributes of perimeter and area of plane shapes, and to make inferences about the effect on both area and perimeter of changing the linear dimensions.

Mathematics:

Math Objects

- [X] Number/Quantity
- [X] Shape/Space
- [ ] Function/Pattern
- [ ] Chance/Data
- [ ] Arrangement

Math Actions (possible weights: 0 through 4)

- [1] Modeling/Formulating
- [2] Transforming/Manipulating
- [2] Drawing Conclusions
- [2] Communicating

Assumed Mathematical Background:

Students should have had instruction on the concepts of area and perimeter and some experience working with geometric shapes.

Core Elements of Performance:

- Correctly compute area and perimeter for the gardens.
- Recognize the shapes that result when the perimeter is reduced and the area increased, and when the perimeter is increased and the area reduced.

Using This Task:

Read through the prompt aloud with your students to ensure that they understand the task.

For informal classroom use, you may adjust the difficulty level of this task by providing more background information to the students. The printed version of the task does not provide any basic facts about area and perimeter, leaving it to the student to recall this information. You might wish to review these topics, including the use of proper units; this considerably reduces the strategic demand of the task.

Extension:

This task may be extended for either instructional or assessment purposes by having students write a prose explanation for the design of their gardens. This considerably increases the communication demand of the task.

Name: _____ Date: _____

GARDENS OF DELIGHT

Here is a map of Ms. McGinty's cucumber garden. In this map, each square corresponds to an area of 1 square yard. The side of each square is 1 yard long.

Ms. McGinty's cucumbers

Name: _____ Date: _____

1. What is the area of Ms. McGinty's cucumber garden?

   What is the perimeter of her garden?

2. Ms. Reilly has planted a cucumber garden with a larger area than Ms. McGinty's garden. However, Ms. Reilly needs **less** fence to enclose her garden. Using the same grid, draw a possible map of Ms. Reilly's garden.

   What is the area of the garden you drew for Ms. Reilly?

   What is the perimeter?

3. Ms. Duffy has planted a cucumber garden with a smaller area than Ms. McGinty's garden. However, Ms. Duffy's needs **more** fence to enclose her garden than Ms. McGinty needs to enclose hers. Again, using the same grid, draw a possible map of Ms. Duffy's garden.

   What is the area of the garden you drew for Ms. Duffy?

   What is the perimeter?

Gardens of Delight

SOLUTION AND RUBRIC

1. One can see, either by counting grid squares or by adding and multiplying dimensions of 3 yards by 8 yards, that the area of Ms. McGinty's cucumber garden is 24 square yards and the perimeter is 22 yards.

2. A possible cucumber garden for Ms. Reilly is a square that is 5 yards on a side. Such a garden would have an area of 25 square yards, but a perimeter of only 20 yards.

3. Ms. Duffy's garden might be long in one dimension but narrow in the other, for instance, 2 yards by 10 yards, which would give a perimeter of 24 yards and an area of 20 square yards. Another possibility is 1 yard by 15 yards, which would give an area of 15 square yards and a perimeter of 32 yards—the greatest perimeter of all the gardens if the dimensions are integers.

	Partial Competency	Full Competency
Modeling/ Formulating *(weight: 1)*	Show some understanding of the difference between perimeter and area.	In each problem, distinguish between the attributes of area and perimeter.
Transforming/ Manipulating *(weight: 2)*	Correctly compute either area **or** perimeter.	Correctly compute area **and** perimeter.
Inferring/Drawing Conclusions *(weight: 2)*	Each part of the question asks for the design of a garden that satisfies two constraints. A student that satisfies only one of the constraints should be awarded partial level.	In **2**, recognize that a garden that is closest to a square is needed to reduce the perimeter but increase the area. In **3**, recognize that an elongated garden is needed to increase the perimeter but reduce the area.
Communicating *(weight: 2)*	State numerical answers clearly **or** Draw clear, correct maps.	Draw clear, correct maps to illustrate the various gardens **and** State all numerical answers clearly, with appropriate units.

4

Pattern and Function

One of the earliest skills that young mathematics students develop is that of recognizing and describing patterns. At the primary level this is evidenced by

- an ability to recognize and generate simple numerical patterns, and to answer questions such as *what is the next number in the sequence 1, 4, 7, 10, 13,...?*
- an ability to recognize and generate spatial patterns, and to answer questions such as *what is the next shape in the pattern X ++ XX +++ ...?*

During the elementary years we expect students to grow in their ability to demonstrate

- enhanced ability to recognize and generate more complex numerical patterns, and to answer questions such as *what is the next number in the sequence 1, 2, 4, 8, ...?*
- enhanced ability to recognize and generate spatial patterns, and to follow directives such as *show two ways to tile a floor with 3-inch square blocks in two different colors so that the pattern is "regular."*

The tasks in this chapter provide the opportunity to demonstrate these abilities in a variety of ways. Tasks involving pattern tend to have a heavy inference demand and may often require solutions to be communicated using words, numbers, and symbols. As students transition into the upper elementary grades, we expect that they will become more facile in describing and generalizing patterns and relationships using algebraic symbolism. They must also become increasingly skilled in the use of correct mathematics vocabulary, facile in modeling and describing relationships, and adept at creating multiple representations of their solutions. One positive outgrowth of the NCTM standards and the use of constructed-response assessments is that young students are being encouraged to visualize the mathematical world in a variety of ways—pictorially, numerically, and verbally. If this becomes a consistent method of problem solving in the elementary grades, students develop habits of mind that will stand them in good stead as they move on to the more complex mathematics of middle school, high school, and college.

WALL DESIGN

TEACHER'S GUIDE

Grade Level: Primary

Description:

This task is designed to assess student ability to identify and extend a geometric pattern.

Mathematics:

Math Objects

- [ ] Number/Quantity
- [X] Shape/Space
- [X] Function/Pattern
- [ ] Chance/Data
- [ ] Arrangement

Math Actions (possible weights: 0 through 4)

- [0] Modeling/Formulating
- [0] Transforming/Manipulating
- [2] Drawing Conclusions
- [2] Communicating

Assumed Mathematical Background:

Students should have some experience with pattern recognition.

Core Elements of Performance:

- Identify and extend a geometric pattern.
- Describe and justify the pattern extension.

Using This Task:

Read through the prompt with your students to ensure that they understand the task. This task may be done in pairs, but each student should provide an individual response.

Students may need to be reminded to shade in their shapes in question 2, so that they exactly match the given pattern.

Kindergarten and early Grade 1 students may have difficulty drawing the objects. You can provide them with stickers that match the given shapes by photocopying the grid onto adhesive paper.

Extension:

Question 2 may well be an extension for some groups. It is also challenging to ask students to write a description of this complex pattern; this substantially increases the communication demand of the task.

Name: _____   **Date:** _____

WALL DESIGN

1. My mother's friend painted my bedroom wall, but she ran out of paint before she finished. Draw in the pictures that are missing on the wall, so that the pattern will stay the same.

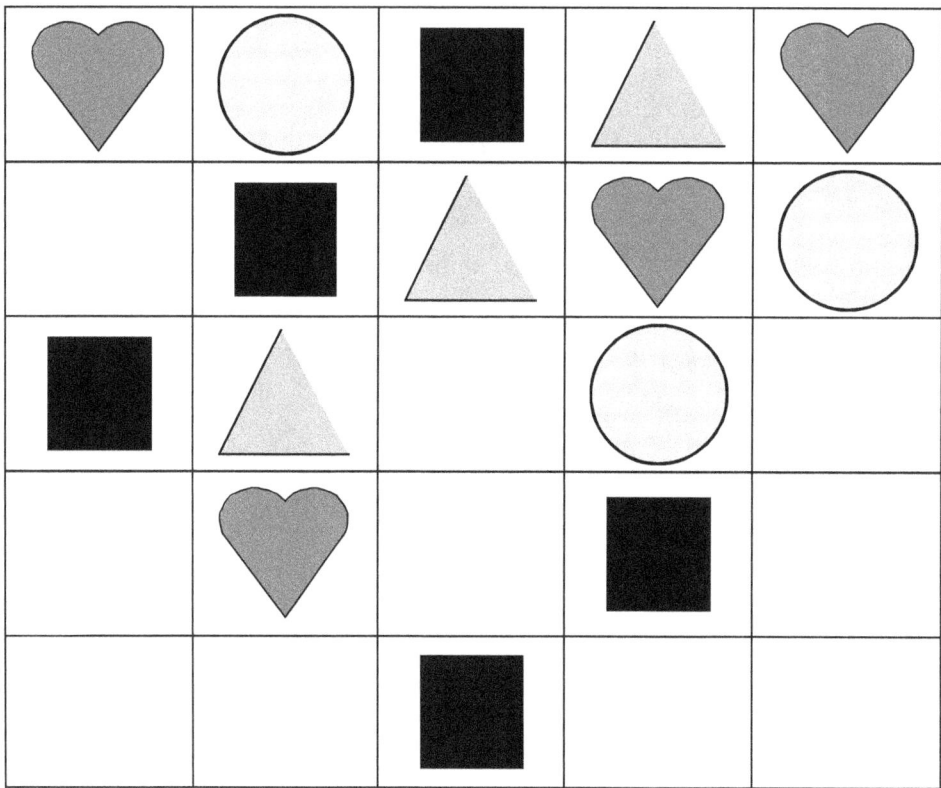

2. How do you know that you are right?

Name: _____ **Date:** _____

3. Here is another wall pattern. Fill in the blank spaces.

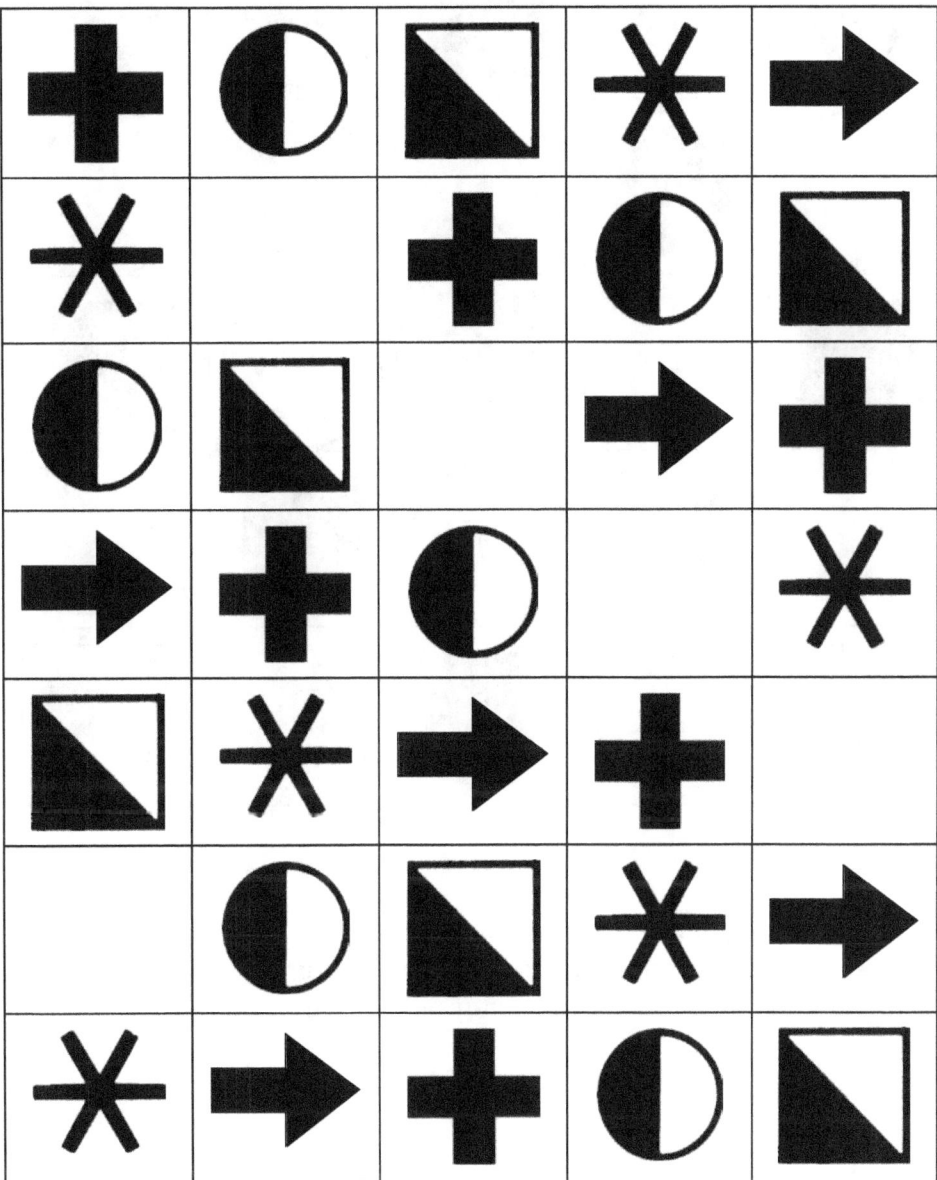

4. How do you know that you are right?

WALL DESIGN

SOLUTION AND RUBRIC

1.

2. Dependent on grade level, students may use phrases such as "I did the crosswise pattern heart, circle, square, triangle" or "The pattern is the same up and down." Older students may notice the repeating diagonal that runs from lower left to top right.

3.

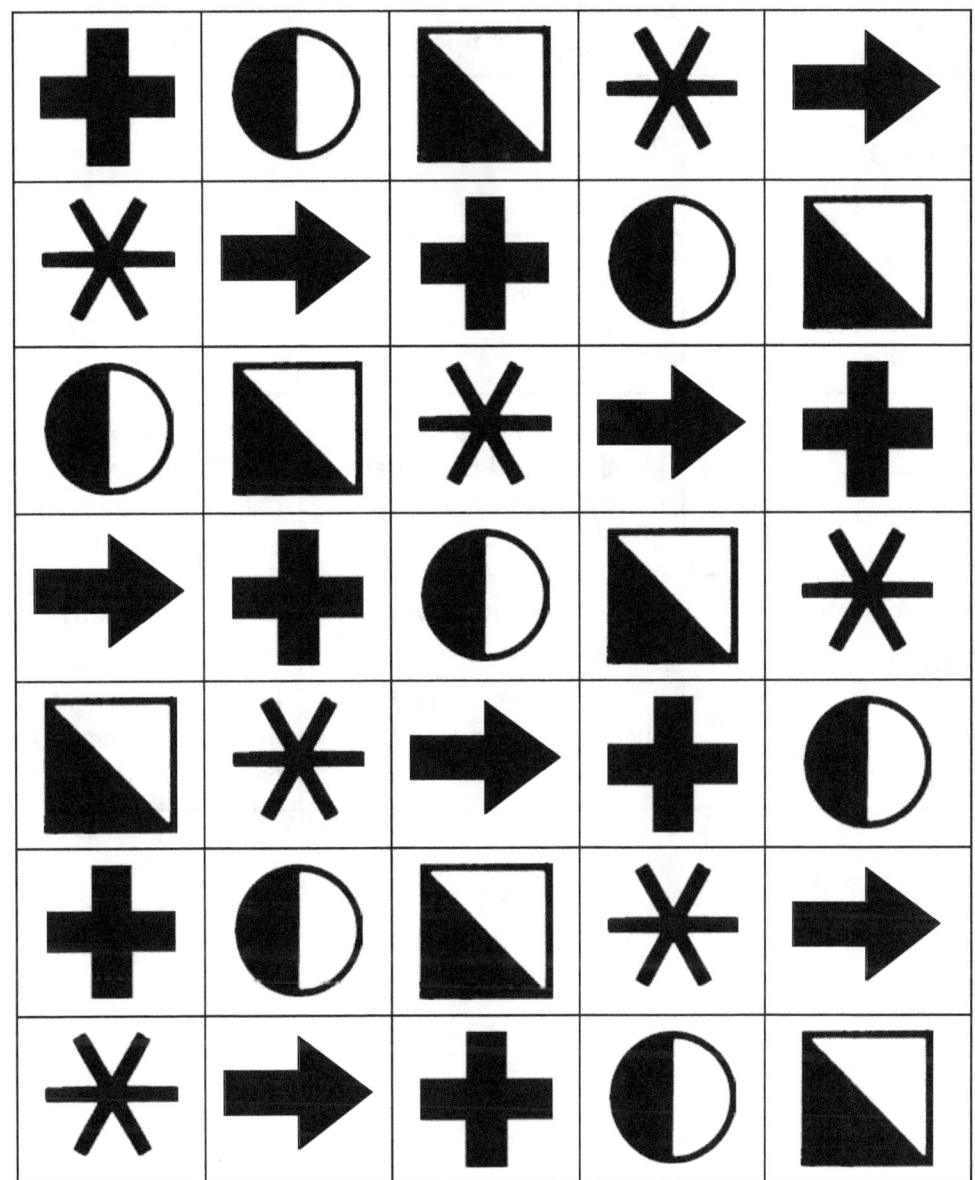

4. Again, dependent on grade level, descriptive responses could be "the up and down pattern is cross, star, circle, arrow, square" or "each crosswise row starts with the fourth shape from the row above" or "every diagonal has the shapes in order."

	Partial Competency	Full Competency
Modeling/ Formulating (weight: 0)		
Transforming/ Manipulating (weight: 0)		
Inferring/Drawing Conclusions (weight: 2)	Student is able to correctly identify extend the pattern for some parts of questions **1** and **2**.	Student is able to correctly identify and extend the pattern for both questions **1** and **2**.
Communicating (weight: 2)	Shapes are not drawn clearly. Prose explanation for questions **2** and **4** are incomplete and/or unclear.	Shapes are clearly drawn and shaded. Prose explanation for questions **2** and **4** are complete and clear.

PATTERN AND FUNCTION   131

Calendar Moves

TEACHER'S GUIDE

Grade Level: Primary/Elementary

Description:

Students are asked various questions about an arrow code that represents moving up, down, left, and right on a calendar grid. They must communicate their answers numerically and verbally.

Mathematics:

Math Objects

[X] Number/Quantity   [ ] Shape/Space   [X] Function/Pattern

[ ] Chance/Data   [ ] Arrangement

Math Actions (possible weights: 0 through 4)

[0] Modeling/Formulating   [1] Transforming/Manipulating

[3] Drawing Conclusions   [3] Communicating

Assumed Mathematical Background:

Students should have worked with monthly calendars and calendar words, and be able to perform simple addition and subtraction.

Core Elements of Performance:

- Determine the meaning of various arrow codes.
- Use direction words, calendar words, and arithmetic words to identify a particular date on a monthly calendar.

Using This Task:

This task is appropriate for both primary and elementary students. It is designed to be used as a portfolio piece or a project that is worked and revised over time. Each question is a relatively free-standing probe. Parts (a) and (b) of questions 1–3 are appropriate for primary students; it should be administered one question at a time on different days. The task is scaffolded so that even upper-elementary students will be challenged by questions 4 and 5.

Extension:

Depending on the grade level at which this task is used, any of the later questions in the sequence may be viewed as an appropriate extension.

Materials Needed:

Monthly calendar

Acknowledgment

This task was inspired by the paper "*Maneuvers on Lattices, an Example of Intermediate Invention*" by David A. Page (1962).

Name: _____ Date: _____

CALENDAR MOVES

There are lots of interesting questions about calendars.

Here is a calendar for a month.

JULY						
Sunday	Monday	Tuesday	Wednesday	Thursday	Friday	Saturday
	1	2	3	4	5	6
7	8	9	10	11	12	13
14	15	16	17	18	19	20
21	22	23	24	25	26	27
28	29	30	31			

Here are some kinds of words that will help you answer questions.

Direction words: left, right, up, down

Calendar words: today, yesterday, tomorrow, last week, this week, next week

Arithmetic words: add, subtract

Name: _____ Date: _____

1. ↓ **Arrows**

    Here are two examples of what a ↓ arrow means on the calendar.

JULY						
Sun.	Mon.	Tue.	Wed.	Thu.	Fri.	Sat.
	1	2	3	4	5	6
7	8	9	10	11	12	13
14	15	16	17	18	19	20
21	22	23	24	25	26	27
28	29	30	31			

    2↓ is 9.

JULY						
Sun.	Mon.	Tue.	Wed.	Thu.	Fri.	Sat.
	1	2	3	4	5	6
7	8	9	10	11	12	13
14	15	16	17	18	19	20
21	22	23	24	25	26	27
28	29	30	31			

    16↓ is 23.

    a. You figure these out:

    3↓ is _____.   11↓ is _____.   24↓ is _____.

Name: _____   Date: _____

- **b.** Using a direction word, tell what the ↓ arrow does.

- **c.** Using a calendar word, tell what the ↓ arrow does.

- **d.** Using an arithmetic word, tell what the ↓ arrow does.

- **e.** What do you think 26↓ should mean? Give a reason for your answer.

- **f.** What else could 26↓ mean?

Name: _____   Date: _____

2. ← **Arrows**

Here is an example of what a ← arrow means on the calendar.

JULY						
Sun.	Mon.	Tue.	Wed.	Thu.	Fri.	Sat.
	1	2	3	4	5	6
7	8	9	10	11	12	13
14	15	16	17	18	19	20
21	22	23	24	25	26	27
28	29	30	31			

19← is 18.

a. You figure these out:

5← is _____.   23← is _____.   30← is _____.

b. Using a direction word, tell what the ← arrow does.

c. Using a calendar word, tell what the ← arrow does.

d. Using an arithmetic word, tell what the ← arrow does.

e. What do you think 14← should mean? Give a reason for your answer.

f. What else could 14← mean?

Name: _____   Date: _____

3. ↑ **Arrows and** → **Arrows**

Can you figure out what is meant by ↑ arrows and → arrows?

a. You figure these out:

23↑ is _____.   27↑ is _____.

18→ is _____.   9→ is _____.

b. Tell what ↑ arrows and → arrows do by filling in the following chart:

	↑ **arrow**	→ **arrow**
What the arrow means, using a direction word		
What the arrow means, using a calendar word		
What the arrow means, using an arithmetic word		

4. Multiple Arrows

You can also do one arrow followed by another arrow.

JULY						
Sun.	Mon.	Tue.	Wed.	Thu.	Fri.	Sat.
	1	2	3	4	5	6
7	8	9	10	11	12	13
14	15	16	17	18	19	20
21	22	23	24	25	26	27
28	29	30	31			

2↓↓ is 16.

JULY						
Sun.	Mon.	Tue.	Wed.	Thu.	Fri.	Sat.
	1	2	3	4	5	6
7	8	9	10	11	12	13
14	15	16	17	18	19	20
21	22	23	24	25	26	27
28	29	30	31			

16↓→ is 24.

a. You figure these out:

22↑↑ is _____.   4→→ is _____.

Name: _____ **Date:** _____

b. Try these:

21↓↑ is _____.   13↑↓ is _____.

Tell what happened with your answers, and explain why it happened.

c. You figure these out:

14↑→ is _____.   6↓← is _____.   26↑← is _____.

d. Try these:

11↓→ is _____.   11→↓ is _____.

Tell what happened with your answers and explain why it happened.

Name: _____   Date: _____

5. Different Calendars

Now let's see if arrows mean the same thing on every calendar.

a. Look back at what you found 23← was on the July calendar in question 2(a).

Now look at a classroom calendar for a different month.

Does 23← turn out to be the same on your classroom calendar? If not, why not?

b. Look back at what 24↓ was on the July calendar. Does it turn out to be the same on your classroom calendar? If not, why not?

c. Make up an arrow problem whose answer on your classroom calendar is different from the answer on the July calendar.

Calendar Moves

Solution and Rubric

1. 3↓ is 10, 11↓ is 18, and 24↓ is 31. The ↓ represents "moving down" or "the same day next week" or "adding 7." This means that 26↓ can be interpreted several different ways. Justifiable answers include 33 (by extending the numerical pattern), 2 (from recognizing that the same day next week will be the 2nd of the following month), or an argument that there is no answer at all because there is no such day on the calendar.

2. 5← is 4, 23← is 22, and 30← is 29. The ← represents "moving left" or "yesterday" or "subtracting 1." This means that 14← can be interpreted several different ways. Justifiable answers include 13 (the preceding day, or the number obtained by subtracting 1), 20 (by wrapping around to the right side of the calendar, but staying on the same row), or an argument that there is no answer at all because the arrow points beyond the edge of the calendar.

3. 23↑ is 16, 27↑ is 20, 18→ is 19, and 9→ is 10. The ↑ represents "moving up" or "the same day last week" or "subtracting 7." The → represents "moving right" or "tomorrow" or "adding 1."

4. 22↑↑ is 8, 4→→ is 6, 21↓↑ is 21, and 13↑↓ is 13. Note that the sequences ↓↑ and ↑↓ just return to the original day (perhaps with exceptions at the edges of the calendar).

    14↑→ is 8, 6↓← is 12, 26↑← is 18, 11↓→ is 19, and 11→↓ is 19. Note that going ↓ and then → leads to the same day as going → and then ↓.

5. Answers will vary depending on the calendar used and also depending on how the arrow operations are interpreted at the edges of the calendar. If the 23rd is a Sunday, there are different ways that 23← could be interpreted (see problem **2**, parts **e** and **f**). If the calendar is for a month with less than 31 days, then there are different ways that 24↓ could be interpreted (see problem **1**, parts **e** and **f**). For any calendar, good answers to **5c** can be created by using days on the edges of the calendar.

	Partial Competency	**Full Competency**
Modeling/ Formulating *(weight: 0)*		
Transforming/ Manipulating *(weight: 1)*	Questions **1(a)**, **2(a)**, **3(a)**, and **4**: Some errors are made. For example, the student may handle single arrows correctly, but have trouble with problems involving two arrows in succession.	Questions **1(a)**, **2(a)**, **3(a)**, and **4**: All arrow problem computations are done correctly, including those involving multiple arrows. No arithmetic errors are made.
Inferring/Drawing Conclusions *(weight: 3)*	Questions **1(e)**, **1(f)**, **2(e)**, **2(f)**: For these arrow problems involving the edge of the calendar, students are able to conceive of only one possible answer (answering **e** but not **f**). Questions **4(b)**, **4(d)**, and **5(c)**: Correct conclusions are reached on some but not all of these problems.	Questions **1(e)**, **1(f)**, **2(e)**, **2(f)**: For these arrow problems involving the edge of the calendar, students are able to conceive of more than one possible answer. Questions **4(b)** and **4(d)**: Correct conclusions are drawn about multi-arrow sequences. Question **5(c)**: The arrow problem invented by the student meets the stated requirement.
Communicating *(weight: 3)*	The full-credit communication criteria are met on some questions but not others. For example, the student might show a weakness in utilizing calendar vocabulary.	Questions **1(b)**, **1(c)**, **1(d)**, **2(b)**, **2(c)**, **2(d)**, and **3(b)**: Students choose appropriate words to describe the behavior of the arrows. Questions **1(e)**, **1(f)**, **2(e)**, **2(f)**, **4(b)**, **4(d)**, and **5**: Reasons and explanations are well-written and complete.

PATTERN AND FUNCTION 143

MAKE A MAP

TEACHER'S GUIDE

Grade Level: Elementary

Description:

This task is designed to demonstrate student ability to read, interpret, and construct diagrams of binary relationships. There is a heavy inference demand.

Mathematics:

Math Objects

☐ Number/Quantity ☐ Shape/Space ☒ Function/Pattern

☐ Chance/Data ☐ Arrangement

Math Actions (possible weights: 0 through 4)

| 2 | Modeling/Formulating | 1 | Transforming/Manipulating |
| 3 | Drawing Conclusions | 2 | Communicating |

Assumed Mathematical Background:

This task assumes basic exposure to reading and interpreting diagrams and using graphic depiction to stand for numerical values or relationships.

Core Elements of Performance:

- Correctly identify all symbols and arrows in the given diagrams.
- Correctly order a group of objects by weight.
- Create a map of a new set of elements and relationships.

Using This Task:

Read through the prompt with your students to ensure that they understand the task.

The pre-activity to this task is essential; it is also important to ascertain that students understand the mathematical meaning of the words "symbol" and "relationship."

Since the task is quite long and has a heavy reading demand, it should be split up for younger students. For example, do the pre-activity and questions 1–3 in one sitting and questions 4 and 5 at another time or as a homework assignment.

Extension:

Question 5 provides an excellent way to extend this task. You can have students revise their original map in order to expand the scope of the relationship described, both pictorially and verbally.

Name: _____ Date: _____

MAKE A MAP

Pre-Activity:

Here is a set of apples: { 🍎 🍎 🍎 🍎 }

There are four apples in this set, and we can represent this set by a black circle: ●

Here is another set of apples: { 🍎 🍎 }

This set is made up of two apples, and we can represent it by a grey circle: ◐

And here is one more apple set: { 🍎 }

We can represent it by a white circle: ○

If I then make a map of the sets of apples and let the arrow say "I am pointing from the set that has more apples to the set that has fewer apples," my map might look like this:

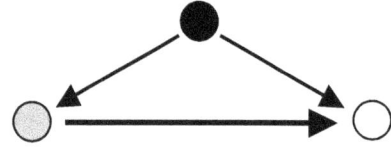

Can you think of another way to draw this map?

Name: _____   **Date:** _____

1. Now let's take sets of numbers and give each one a different symbol so that you can tell them apart:

   {5 6 7 8} ⬜

   {5 6 8}   △

   {6 8}   ⊕

   {5 7}   ⊗

   The arrows say "I am pointing from the set that has more numbers in it to the set that has fewer numbers in it."

   Look at the drawing below, which shows the relationships between the four number sets. Unfortunately, all of the arrows haven't been drawn correctly. How would you correct this map?

   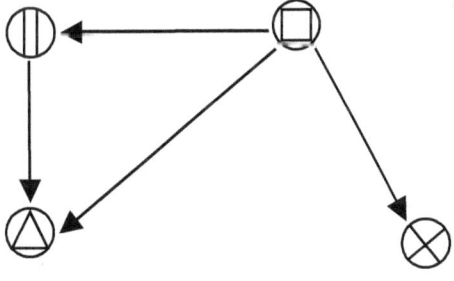

2. Kevin has visited Andrew, and Michael has also visited Andrew, but Kevin and Michael haven't visited each other. Kevin is older than Andrew and Michael, who are the same age.

The relationships between these friends are shown in the maps below. The solid arrow says "I have visited him." The dotted arrow says "I am older than him." In these drawings, which triangle stands for Kevin, which stands for Andrew, and which stands for Michael?

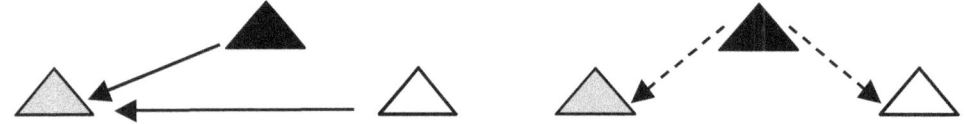

3. Beth and Maria live in the same house. Maria and Amy go to the same school. Amy and Beth go to the same camp.

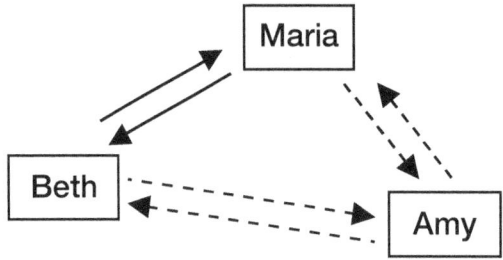

Look at the map and tell which arrows say "she and I live in the same house."

Name: _____ Date: _____

What do the dotted arrows at the bottom of the map say?

How about the third set of arrows?

4. A portable sewing machine weighs 12 pounds, a radio and a vacuum cleaner each weigh 9 pounds, an iron weighs 3 pounds, and a hair dryer weighs 2 pounds. In the drawing below, each circle represents an object, and each arrow points from a lighter object to a heavier object.

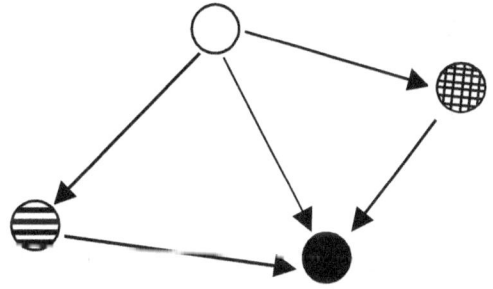

a. Which circle could stand for the iron?

b. What do the other circles stand for?

c. Can you be sure what each circle stands for? Why or why not?

Name: _____   **Date:** _____

5. Now think of a relationship among three people or things, and give each one a symbol.

   a. Draw a map of the relationship, using the symbols you chose and arrows that point in the correct way. Be sure to tell what your symbols and arrows stand for.

   b. Tell in words the story of the relationship you have mapped.

Make a Map

SOLUTION AND RUBRIC

PATTERN AND FUNCTION 149

1.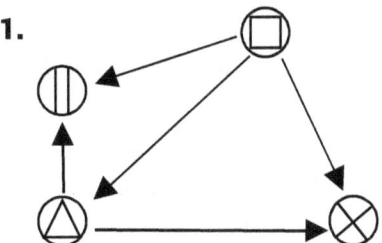

2. Kevin = black triangle; Andrew = gray triangle; Michael = white triangle.

3. Solid line with arrow = we live in the same house; dashed line with arrow = we go to the same school; dotted line with arrow = we go to the same camp.

4. The only circle that can be definitely identified is the black, representing the sewing machine; for the other circles, there are several possibilities.

   Here is one:

   The black circle represents the sewing machine; the first shaded circle represents the radio; the second shaded circle represents the vacuum cleaner; the white circle represents either the iron or the hair dryer.

   And here is another:

   The black circle represents the sewing machine; the first shaded circle represents the radio or the vacuum cleaner; the second shaded circle represents the iron; the white circle represents the hair dryer.

   Dealing with uncertainty is a hard task for this age group, and they may have difficulty verbalizing the possible duality of representation.

5. It is imperative that the symbolism, the narrative of the relationships, and the direction of the arrows be consistent. It is hoped that students will choose interesting, fanciful stories to map. Here is one possibility:

   There are 3 surviving kittens in a litter. There is a female named Anne, who weighs the least and is the palest. There is a male named Charlie who weighs the most, and there is a female named Beth who has the darkest fur.

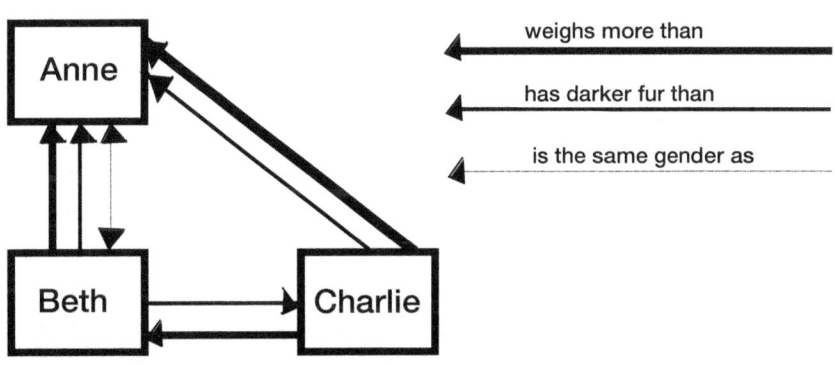

	Partial Competency	**Full Competency**
Modeling/ Formulating *(weight: 2)*	Correctly identify some of the symbols and arrows in **1–4**.	Correctly identify all of the symbols and arrows in **1–4**.
Transforming/ Manipulating *(weight: 1)*	Give a partial ordering of the objects in **4**.	Make a fully correct ordering of the objects in **4**.
Inferring/Drawing Conclusions *(weight: 3)*	Correctly identify elements and arrows in **1–3**.	Additionally, provide a reasonable interpretation of **4**.
Communicating *(weight: 2)*	Give a minimal narrative in **5** that mimics one of the earlier questions.	Devise and describe clearly a fresh set of elements and relationships in **5**.

Coding the Alphabet

TEACHER'S GUIDE

Grade Level: Elementary

Description:

Students are asked to demonstrate their ability to perform simple calculations, as well as to weigh the effects of multiple solutions to a problem.

Mathematics:

Math Objects

- [ ] Number/Quantity
- [ ] Shape/Space
- [x] Function/Pattern
- [ ] Chance/Data
- [ ] Arrangement

Math Actions (possible weights: 0 through 4)

- [0] Modeling/Formulating
- [2] Transforming/Manipulating
- [2] Drawing Conclusions
- [1] Communicating

Assumed Mathematical Background:

Basic addition is required for the first part of this task; the second part requires experience with using logical reasoning to solve problems.

Core Elements of Performance:

- Correctly "translate" from letters to numbers and calculate values using addition.
- Make a systematic listing of possible number combinations.
- Recognize that adding constraints to a particular situation leads to a unique solution.

Using This Task:

Read through the prompt with your students to ensure that they understand the task. If students have difficulty adding horizontally after they have assigned the numerical values in question 1(a), suggest that they rewrite the sums and add vertically.

Encourage students to use a dictionary to check the spelling of the animal names in question 1(b). Unless this task is being used as an assessment of student ability to perform multi-digit addition, they should be allowed to use calculators for this part of the task.

It may be necessary to have an open discussion about how you can be sure you have found all the possible sets of numerical values in question 2(a).

Extension:

Question 2(b) is quite challenging and should be considered an extension question for some students.

Materials Needed:

Dictionaries, calculators

Name: _____   Date: _____

CODING THE ALPHABET

Suppose we assign a value to every letter in the alphabet:

A	B	C	D	E	F	G	H	I	J	K	L	M	N	O	P	Q	R	S	T	U	V	W	X	Y	Z
1	2	3	4	5	6	7	8	9	10	11	12	13	14	15	16	17	18	19	20	21	22	23	24	25	26

The value of a word is found by changing its letters into numbers and adding them.

For example, the word BAD has the value of 7 because 2 + 1 + 4 = 7.

1. **a.** Write the value of each letter beside the words listed below, and find the values of the following animal names:

    cow

    dog

    pig

    lion

    deer

    bear

    b. Find an example of an animal whose name's value is between 100 and 200.

Name: _____ **Date:** _____

2. **a.** Joan has decided to change the code and has given every letter some secret number. If you know that the value of **ON** is 5 and the value of **TO** is 7, what can **O**, **N**, and **T** be?

Can you find another set of number values for **O**, **N**, and **T**?

Are there other sets of number values? How do you know you have found them all?

b. Using this **same** code, how many different solutions for **O**, **N**, and **T** will you have if you are given the additional information that the value of **TOO** is 9?

CODING THE ALPHABET — SOLUTION AND RUBRIC

1a. {COW} = 41   {DOG} = 26   {PIG} = 32   {LION} = 50
{DEER} = 32   {BEAR} = 26

b. One must find an animal name that is fairly long and that has some high-valued letters. Examples are {RHINOCEROS} = 124 and {HIPPOPOTAMUS} = 169.

2a. There are six possible values for each letter, as shown on the following table.

O	N	T
0	5	7
1	4	6
2	3	5
3	2	4
4	1	3
5	0	2

2b. Since O must have the same value in each use, there is only one solution that gives nine as the value of TOO: O = 2, N = 3, T = 5. This is a unique solution.

	Partial Competency	Full Competency
Modeling/ Formulating *(weight: 0)*		
Transforming/ Manipulating *(weight: 2)*	Correctly calculate the values of the given animal names.	Additionally, find an animal name with a value between 100 and 200, and correctly calculate this value.
Inferring/Drawing Conclusions *(weight: 2)*	Find some of the possible solutions for 2a.	Find all solutions to 2a and recognize that adding an additional requirement leads to a unique solution of the problem.

5

Chance and Data

In the primary grades, students have an intuitive sense of "counting up" elements of their world. We expect them to grow in their ability to

- collect, organize, and display simple data sets, and to respond to directives such as *make a presentation of all the kinds of pets owned by the students in your class;*
- make decisions based on provided data, and to answer questions such as *given a temperature chart, which months are best to grow vegetables?*

As students move through the elementary grades (3–5) and then to upper elementary and transition (6–7), we expect them to improve their facility to

- collect, organize, and display data sets using multiple presentation devices, and to respond to directives such as *make a presentation of all four-legged pets owned by the students in your class. Include their weights, ages, and length from nose to tip of tail;*
- use provided data to extrapolate additional information, and to answer questions such as *given a bar graph that represents the batting averages of the sixth grade baseball players by team, which team do you think will win the championship?*

Tasks that focus on data tend to have a high communication demand; they also may require substantial modeling and transforming of the original information in order to aggregate or represent it. These aspects are reflected in the weighting of the mathematical actions in this chapter.

Beach Day

TEACHER'S GUIDE

Grade Level: Primary

Description:

This task assesses student ability to interpret data presented in a weather chart and to justify decisions that are based on the data.

Mathematics:

Math Objects

	Number/Quantity		Shape/Space		Function/Pattern
X	Chance/Data		Arrangement		

Math Actions (possible weights: 0 through 4)

0	Modeling/Formulating	0	Transforming/Manipulating
3	Drawing Conclusions	3	Communicating

Assumed Mathematical Background:

This task assumes some familiarity with reading weather charts.

Core Elements of Performance:

- Interpret the data presented in the given chart.
- Make a decision based on this data.
- Explain why other options will not work.

Using This Task:

Teachers should read the prompt to the students. In an informal classroom situation, it may be useful to emphasize that the students are to use the information in both the story **and** the chart to make their decision. They may also need to be reminded to tell why they did not choose **each** of the other days.

Extension:

Ask students to determine the differences in temperature from one day to the next—this provides good practice in double-digit subtraction, in some cases involving regrouping. It also provides an appropriate context in which to talk about positive and negative numbers.

Name: _____ Date: _____

BEACH DAY

The Smith family is planning to make a day trip to the beach sometime this week, but is not sure which day would be the best.

Willy Smith has a soccer game on Tuesday, and his sister is going to get an award for Math Camp on Friday.

Below is a chart of the weather forecast for this week.

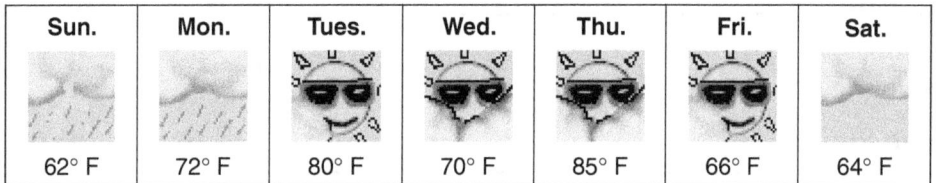

Sun.	Mon.	Tues.	Wed.	Thu.	Fri.	Sat.
62° F	72° F	80° F	70° F	85° F	66° F	64° F

1. Using the story and the chart, choose the best day for the Smith family to spend at the beach. Explain why you chose that day, including a description of the weather.

2. Explain your reasons for NOT choosing the other days.

Beach Day

SOLUTION AND RUBRIC

1. Most students will pick Thursday because there are no other activities going on, and the temperature is the highest of any of the remaining days. However, some students may suggest going to the beach after the soccer game on Tuesday, as this is the hottest, sunniest day; this is a less appropriate answer.

2. Tuesday and Friday are explicitly eliminated because of the activities mentioned in the story. Sunday, Monday, and Saturday are eliminated because of the rainy or cloudy weather and low temperatures shown on the weather chart. Wednesday and Thursday have the same amount of sun, but the temperature on Wednesday is considerably lower than on Thursday.

	Partial Competency	Full Competency
Modeling/ Formulating *(weight: 0)*		
Transforming/ Manipulating *(weight: 0)*		
Inferring/Drawing Conclusions *(weight: 3)*	Student is able to interpret some of the given data in order to make a decision on a suitable day.	Student uses all of the given data to make a decision, using both the story and the weather chart. There is careful consideration of why the other days will not work.
Communicating *(weight: 3)*	Student does not give justification for choice in question **1** and/or does not explain why each of the other days was not suitable in question **2**.	Student gives a complete justification for choosing a certain day, including a description of the weather. The unsuitability of each of the other days is described clearly and completely.

We Scream for Ice Cream

TEACHER'S GUIDE

Grade Level: Primary

Description:

This task assesses student ability to interpret data from a pictograph and to perform basic arithmetic manipulations.

Mathematics:

Math Objects

- [X] Number/Quantity
- [ ] Shape/Space
- [ ] Function/Pattern
- [X] Chance/Data
- [ ] Arrangement

Math Actions (possible weights: 0 through 4)

- [0] Modeling/Formulating
- [1] Transforming/Manipulating
- [2] Drawing Conclusions
- [3] Communicating

Assumed Mathematical Background:

This task assumes some familiarity with pictographs.

Core Elements of Performance:

- Interpret the data presented in the pictograph.
- Perform arithmetic calculations correctly.

Using This Task:

Read through the prompt with your students. In an informal classroom situation, it may be useful to emphasize that students should base their answers only on the information in the pictograph, not on their own flavor preferences!

Extension:

Question 4 will be an extension for all but the strongest Grade 2 students.

Students can be asked to survey their class for favorite flavors and collect the data using tally marks; older students can then use the data to make a different type of graph, such as a bar graph.

Name: _____ Date: _____

WE SCREAM FOR ICE CREAM

Students in Mr. Max's class are planning an ice cream party. Each student voted for their favorite flavor, and the students made a chart of the results:

Chocolate	🍦	🍦	🍦	🍦	🍦	🍦		
Vanilla	🍦	🍦	🍦	🍦	🍦	🍦	🍦	🍦
Strawberry	🍦	🍦	🍦					
Orange Sherbet	🍦	🍦	🍦	🍦	🍦	🍦	🍦	

1. How many students are in Mr. Max's class? Explain how you know this.

Name: _____ **Date:** _____

2. If the class can buy only two different ice cream flavors for the party, which two should they buy? Why?

3. How many students will not get their favorite flavor? Explain how you know this.

Name: _____ **Date:** _____

Extension:

Chocolate	🍦	🍦	🍦	🍦	🍦	🍦		
Vanilla	🍦	🍦	🍦	🍦	🍦	🍦	🍦	🍦
Strawberry	🍦	🍦	🍦					
Orange Sherbet	🍦	🍦	🍦	🍦	🍦	🍦	🍦	

4. Suppose that this chart represents the votes from a class where each student voted for their two favorite flavors. How many students are in this class? Explain how you know this.

CHANCE AND DATA

We Scream for Ice Cream

SOLUTION AND RUBRIC

1. There are 24 students in Mr. Max's class. Students may explain this answer by saying "I counted the ice cream cones" or, occasionally, "I counted all the squares and then subtracted the empty ones."

2. They should buy vanilla and orange sherbet since these two flavors received the most votes. Some students will say chocolate rather than orange sherbet "because I like chocolate the best"—this is not an appropriate answer!

3. Nine students will not get their favorite flavor. This may be expressed as "all the people who voted for chocolate and strawberry" or "24 (total students) – 15 (vanilla and orange sherbet voters)."

Extension:

4. Students will have varying ways of explaining the answer that there are twelve students in this class. Some may use their answer for question 1 and say that there are only half as many in this class, because there are two cones for each student. Others may take a less efficient, but equally correct route of grouping the cones in groups of two and then counting up the groups.

	Partial Competency	Full Competency
Modeling/ Formulating *(weight: 0)*		
Transforming/ Manipulating *(weight: 1)*	Student performs some of the required additions and/or subtractions correctly.	Student performs all of the required additions and subtractions correctly.
Inferring/Drawing Conclusions *(weight: 2)*	Student is able to interpret some of the data presented in the pictograph and arrives at a reasonable conclusion for questions **1**, **2**, or **3**.	Student is able to interpret all of the data presented in the pictograph and arrives at reasonable conclusions for questions **1**, **2**, and **3**.
Communicating *(weight: 3)*	Explanations for the numerical answers in questions **1**, **2**, and **3** are unclear, incomplete, or incorrect.	Explanations for all numerical answers are clear and complete, and make use of appropriate mathematics vocabulary and comparison words.

MIXED-UP SOCKS

TEACHER'S GUIDE

Grade Level: Elementary

Description:

This task assesses student ability to reason logically in a real-world situation modeled by discrete mathematics.

Mathematics:

Math Objects

☐ Number/Quantity ☐ Shape/Space ☐ Function/Pattern

[X] Chance/Data [X] Arrangement

Math Actions (possible weights: 0 through 4)

[2] Modeling/Formulating [1] Transforming/Manipulating

[2] Drawing Conclusions [2] Communicating

Assumed Mathematical Background:

This task assumes a basic understanding of simple probability.

Core Elements of Performance:

- Determine the fewest number of selections that will assure a certain outcome.
- Be able to extend the basic concept to more complex situations.

Using This Task:

Read through the prompt with your students to ensure that they understand the task. The printed version of the task does not provide students with any basic information about probability; for informal classroom use, whether instructional or diagnostic, you can adjust the difficulty level by providing more instruction. Most students will need to have the concept of "being certain" explored.

The pre-activity is essential and should not be omitted.

It is very helpful to have some form of manipulatives available in order to demonstrate the possibilities to the students.

Extension:

It is important that this not be viewed as a "probability without replacement" task, but rather as a situation of considering worst-case scenarios. However, more advanced students may be asked to think and write about whether the probability of getting a certain color changes once you have pulled a sock of that color.

Materials Needed:

Red, blue, and green chips or pieces of colored paper

Name: _____   Date: _____

Mixed-Up Socks

Pre-Activity:

6 brown socks and 4 white socks are all mixed up in a dresser drawer. The 10 socks are exactly alike except for their color. The room is in total darkness, and you want two matching socks.

Start pulling socks out of the drawer, one at a time. What is the fewest number of socks that you must take out of the drawer in order to be certain that you have a pair that matches? Explain your answer.

Name: _____ Date: _____

Task:

12 red socks and 8 blue socks are all mixed up in a dresser drawer. The 20 socks are exactly alike except for their color. The room is in total darkness, and you want two matching socks.

1. Start pulling socks out of the drawer, one at a time. What is the fewest number of socks that you must take out of the drawer in order to be certain that you have a pair that matches?

2. What is the smallest number of socks that you must take out of the drawer in order to be certain that you have a pair of red socks?

_____

3. What is the smallest number of socks that you must take out of the drawer in order to be certain that you have two pairs of matching socks (two pairs of red socks or two pairs of blue socks) or one pair of red socks and one pair of blue socks? Explain your answer.

4. You add 6 green socks to the 12 red and 8 blue socks that are already in the drawer. Now what is the least number of socks that you must take out of the drawer in order to be certain that you have a pair that match? Explain your answer.

Mixed-Up Socks — SOLUTION AND RUBRIC

Pre-activity:

Certainly, taking two socks out of the drawer is not enough to guarantee a pair, since you might get one brown and one white sock; but the third sock drawn, whether brown or white, is guaranteed to create a match.

Task:

1. The situation is exactly the same as the pre-activity and is independent of the number of socks of each color; so, three socks will guarantee a match.

2. If seeking a red pair, the worst case scenario is that all eight blue socks are drawn before you get two reds. Thus, it is necessary to draw ten socks to be sure of getting a red pair.

3. If you begin by drawing three socks out of the drawer, there is guaranteed to be a pair (see question 1) plus one extra sock. Again, by the results in 1, taking the leftover sock and drawing two more socks guarantees another pair; so, by drawing five socks, you are guaranteed to have two matched pairs.

4. This question calls for a generalization of the results of 1. With the addition of the third color, drawing three socks is not enough, as you might get one red, one blue, and one green. By taking a fourth sock, you are sure to have a match.

	Partial Competency	Full Competency
Modeling/ Formulating *(weight: 2)*	Devise an appropriate counting scheme for any two of questions **1–3**.	Devise an appropriate counting scheme for questions **1–3** and correctly generalize to the three-color situation in **4**.
Transforming/ Manipulating *(weight: 1)*	Arrive at correct numerical answers for some of the questions.	Correctly carry out the counting schemes to arrive at correct numerical answers for all questions.
Inferring/Drawing Conclusions *(weight: 2)*	Use the results from **1** to solve the more general problem in **4**.	Successfully apply the results from **1** and **2** to the more complex situations in **3** and **4**.
Communicating *(weight: 2)*	Communicate the numerical answers clearly and give a limited explanation in **3** and **4**.	Communicate numerical answers clearly and display evidence that supports the answers to **3** and **4**.

Measure Me

TEACHER'S GUIDE

Grade Level: Elementary

Description:

This task assesses student ability to read, interpret, and construct scatter plots, and to explore the relationships among data points.

Mathematics:

Math Objects

- [ ] Number/Quantity
- [ ] Shape/Space
- [X] Function/Pattern
- [X] Chance/Data
- [ ] Arrangement

Math Actions (possible weights: 0 through 4)

- [2] Modeling/Formulating
- [1] Transforming/Manipulating
- [2] Drawing Conclusions
- [2] Communicating

Assumed Mathematical Background:

This task assumes a basic understanding of how data is represented in a scatter plot.

Core Elements of Performance:

- Give reasonable assumptions and correctly interpret points on a scatter plot.
- Construct a scatter plot using appropriate labels and scale.

Using This Task:

Read through the prompt with your students to ensure that they understand the task. The first two questions can be done individually, but it is essential to do the third question in small groups of three or four.

Students may need to be reminded that their answers to questions 1 and 2 should reflect only information that is available from the graph.

Extension:

Question 4 will be an extension for students that have difficulty composing a verbal response without having a specific graphic representation to describe.

Materials Needed:

Yard/meter sticks, tape measures, grid paper (optional)

Measure Me

As a class project, a kindergarten class measured the height of the students in the class. They also weighed each student. Here is what these measurements looked like when they drew them on a grid.

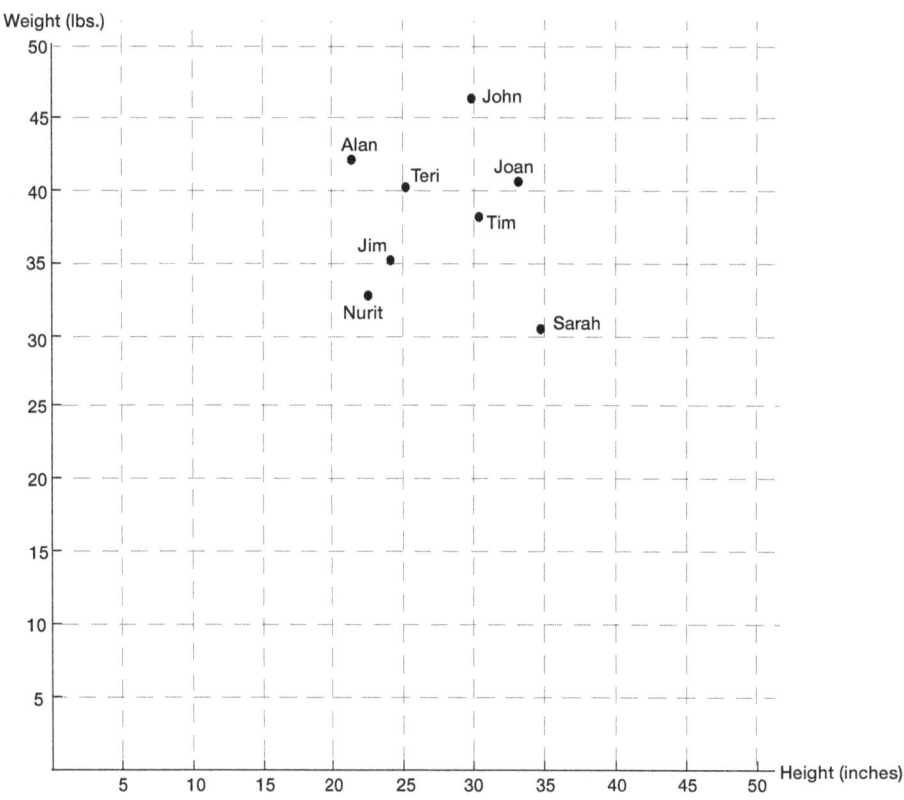

1. List at least three things you can tell about the children in this kindergarten class from looking at this graph.

Name: _____   Date: _____

2. a. If your class were to do a similar project, list at least two ways you think your class graph would be different from the kindergarten picture.

b. In what ways do you think your picture would be similar to the kindergarten picture?

3. Measure the height of each person in your group, and the length of their hand from the tip of the middle finger to the wrist. Make a list of this Information.

Name: _____ Date: _____

Now use the grid to draw the graph for your group.

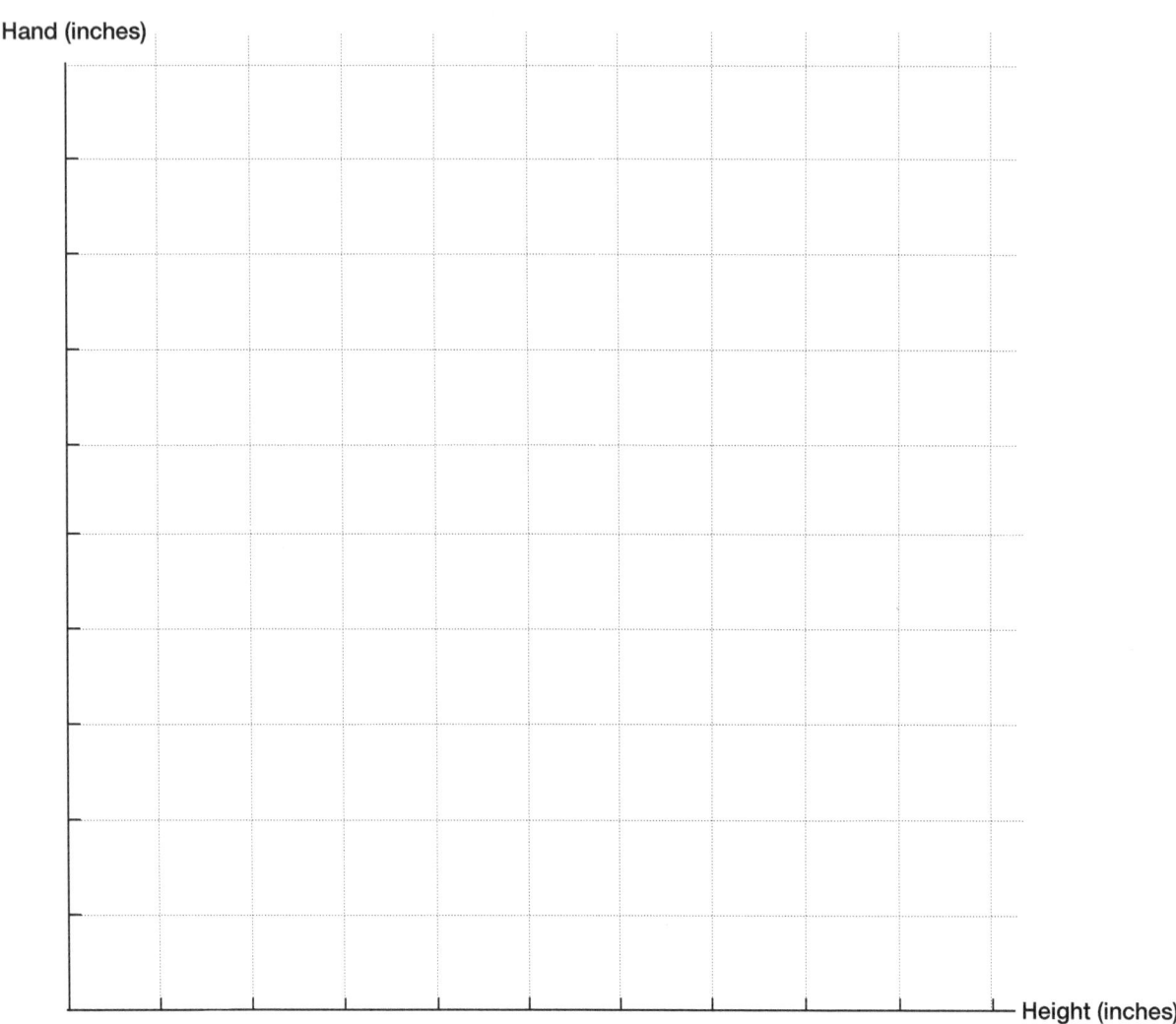

Extension:

4. If you were to do a similar project with 100 people at a baseball game, describe in words what you think the diagram would look like.

MEASURE ME SOLUTION AND RUBRIC

1. Some possible responses are: John is the heaviest, Sarah weighs the least, Sarah is the tallest, Alan is the shortest boy, Nurit is the lightest and the shortest girl, Sarah is both very tall and weighs the least so she must be pretty slim and so forth.

2. **a.** Fourth graders are generally taller and heavier than kindergarten students, so the set of points would be moved diagonally towards greater values of both coordinates.

   b. Although there will be some differences or "scatter" between heights and weights in the fourth grade, all of these points will stay in a relatively small area on the grid, and the spread between the points will not be noticeably different from the spread between the kindergarten points.

3. Again, there will be some difference between heights and hand length in fourth grade students, but all of the points should stay in a relatively small area on the grid.

4. Since people coming to a baseball game are of different ages, the scatter between the points on the diagram will be substantially greater than in the previous cases.

	Partial Competency	Full Competency
Modeling/ Formulating *(weight: 2)*	Correctly place points on the scatter plot taking **either** height **or** hand length into account.	Correctly place points on the scatter plot taking **both** attributes into account.
Transforming/ Manipulating *(weight: 1)*	Draw the picture of the class in **3** with partial accuracy.	Correctly draw the group graph in **3** and have numerical labeling reasonable and consistent with units.
Inferring/Drawing Conclusions *(weight: 2)*	Give reasonable assumptions and correctly interpret points on the scatter plot in **1** and **2**.	Additionally, give reasonable assumptions in **4**.
Communicating *(weight: 2)*	Neglect to include reasonable labels or to make a list of the measurements in **3**. Provide an unclear or incomplete explanation in **2a**, **2b**, and **4**.	Draw a clear grid in **3** accompanied by a list of the measurements and use appropriate labels on the axis. Provide a full verbal explanation in **2a**, **2b**, and **4**.

6

Arrangement

In the primary and early elementary grades, the mathematical objects we refer to as arrangements tend to be closely connected to the idea of repetitive pattern. However, as students begin to explore more complicated enumerations and to organize and represent larger amounts of discrete information, subtle differences show up in the strategies needed to solve the problems. Consequently, when younger students encounter questions such as *How many different ways can you seat four people at a square table so that there is one person on each side of the table?*, they may see it, and solve it, as a repetitive pattern situation. As they move through the curriculum, however, they must become increasingly able to enumerate and organize more complex combinations and permutations.

Therefore, by the time students have reached the upper elementary and transition grades (5–7), it is important that they have developed

- orderly, efficient methods for keeping track of the enumerations, and ability to answer questions such as *In a clothing store, you have a choice of 20 different kinds of T-shirts and 5 different kinds of shorts. How many different outfits can you put together from this selection?*
- ability to aggregate and display solutions with precision.

The tasks in this chapter vary slightly in the demands that they make with respect to the four mathematical actions of Modeling/Formulating, Transforming/Manipulating, Inferring/Drawing Conclusions, and Communicating, but the combined weightings of each action are relatively equal and developmentally appropriate. Students must display inventiveness in bringing together disparate elements of what they know in order to solve the problem; therefore, there will often be more than one correct approach and/or answer.

When working arrangement tasks, it is particularly important to have manipulatives available so that students can physically model their understanding of the situation. The ability to translate from the physical to the conceptual is a skill that requires constant nurturing in the primary and elementary years.

ARRANGEMENT 177

Postal Puzzles

TEACHER'S GUIDE

Grade Level: Primary

Description:

This task assesses student ability to investigate the properties of multiple combinations of two given numbers.

Mathematics:

Math Objects

- [X] Number/Quantity
- [ ] Shape/Space
- [ ] Function/Pattern
- [ ] Chance/Data
- [X] Arrangement

Math Actions (possible weights: 0 through 4)

- [0] Modeling/Formulating
- [2] Transforming/Manipulating
- [2] Drawing Conclusions
- [1] Communicating

Assumed Mathematical Background:

Students should know basic number facts.

Core Elements of Performance:

- Determine the total cost for a combination of four stamps of given values.
- Determine least and greatest cost.
- Find all possible combinations of the given stamps.

Using This Task:

Read through the prompt with your students to ensure that they understand the task.

It is important to spend sufficient time on the provided pre-activity in order for students to fully understand the context of the questions. With younger children, it may be helpful to have them manipulate stamps (stickers) printed with the proper value or colored chips that are assigned the given values. Kindergarten and Grade 1 students will need assistance in organizing their data presentation for question 4; here again, it will be useful to have some sort of manipulative to represent the stamps.

Extension:

Students may be asked to write a prose explanation of how they know that they have found all the possible arrangements. This adds significantly to the communication demand of the task.

Materials Needed:

Stickers, colored chips, or tiles

Name: _____ Date: _____

Postal Puzzles

Pre-Activity:

Your classmate is holding a letter that has two stamps on it. She says that either stamp could cost 4 cents or 7 cents.

    a. What is the largest amount of postage that might be on the letter?

    b. What is the smallest amount of postage that might be on the letter?

    c. Are there any other amounts of postage that might be on the letter?

Name: _____   Date: _____

Task:

In the imaginary country of Philalia, the only kinds of stamps are those that cost 3 tekos and those that cost 5 tekos. In Philalia, every item mailed must have exactly 4 stamps on it.

1. Write one possible combination of stamps you could use in Philalia. What is its total cost?

2. What is the least amount it could cost to send something in Philalia?

3. What is the greatest amount it could cost?

4. Write all the possible ways you can combine stamps.

   How much does each combination cost?

Postal Puzzles — SOLUTION AND RUBRIC

1. Any combination of four stamps, either all value 3, all value 5, or a mix of 3 and 5 is acceptable. The complete list is given in the answer to question 4.

2. The least value is achieved if each of the stamps has the lowest possible value. Since the value of each stamp is 3 tekos or 5 tekos, the lowest value is 3. Therefore, the least amount of postage is obtained when each stamp is a 3-teko stamp: $3 + 3 + 3 + 3 = 12$.

3. Similarly, the greatest amount is achieved when each stamp has its greatest possible value, that is 5 tekos: $5 + 5 + 5 + 5 = 20$.

4. While it is possible to make a long list of possible combinations and then eliminate duplicates, it is much easier to make such a list by finding some organizing properties. Since the order of the stamps is not important (the total postage does not change as stamps are rearranged), the only factor is the number of 3-teko and the number of 5-teko stamps. Since the total number of stamps is four, knowing the number of one kind of stamp automatically gives the number of the other kind, and the total number of either stamp can be 0, 1, 2, 3, or 4.

   The totals are $5 + 5 + 5 + 5 = 20$; $5 + 5 + 5 + 3 = 18$; $5 + 5 + 3 + 3 = 16$; $5 + 3 + 3 + 3 = 14$; $3 + 3 + 3 + 3 = 12$.

	Partial Competency	Full Competency
Modeling/ Formulating *(weight: 0)*		
Transforming/ Manipulating *(weight: 2)*	Student correctly computes some of the postages and/or correctly identifies the least or the greatest postage amounts.	Student completes all the computations correctly, and identifies the least and the greatest postage amounts that match the list in question **4**.
Inferring/Drawing Conclusions *(weight: 2)*	Student provides partial reasoning in either question **2** or **3** and/or makes a list that is not complete, organized, or systematic.	Student completes the list in question **4** in a systematic manner, verifying the completeness of the list and providing reasoning for questions **2** and **3**.
Communicating *(weight: 1)*	Student makes a partial list of stamp combinations, and/or gives a stamp combination only in question **1**.	Student makes a full list of possible stamp combinations.

Shares

Grade Level: Primary

TEACHER'S GUIDE

Description:

This task assesses student ability to partition a number into all possible number pairs.

Mathematics:

Math Objects

| [X] Number/Quantity | [ ] Shape/Space | [ ] Function/Pattern |
| [ ] Chance/Data | [X] Arrangement | |

Math Actions (possible weights: 0 through 4)

| [2] Modeling/Formulating | [1] Transforming/Manipulating |
| [0] Drawing Conclusions | [1] Communicating |

Assumed Mathematical Background:

Students should know basic number facts.

Core Elements of Performance:

- Determine all possible number pairs.

Using This Task:

Read through the prompt with your students to ensure that they understand the task. Using a counting manipulative to model the problem will help younger students.

Extension:

Students can be asked to provide a prose explanation of how they know that they have found all the possible combinations; this substantially increases the communication demand of the task.

Materials:

Any counting manipulative, such as tiles, plastic chips and so forth.

Name: _____   Date: _____

SHARES

Josephine and her brother Paul have 9 marbles together. They can share them in different ways. For example, Josephine could take all the marbles, or Josephine could take 5 and Paul could take 4.

Fill in the chart with all the different ways that Josephine and Paul could share the marbles.

Josephine	9									
Paul	0									

SHARES

SOLUTION AND RUBRIC

The combinations should all have a combined total of 9 marbles. Complete solutions should include organized tables with pairs of numbers in decreasing order for Josephine, from 9 to 0.

| Josephine | 9 | 8 | 7 | 6 | 5 | 4 | 3 | 2 | 1 | 0 |
| Paul | 0 | 1 | 2 | 3 | 4 | 5 | 6 | 7 | 8 | 9 |

	Partial Competency	Full Competency
Modeling/ Formulating *(weight: 2)*	Student gives a partial and disorganized list of pairs of marbles.	Student completes the table in a systematic manner.
Transforming/ Manipulating *(weight: 1)*	Some number pairs are incorrect.	Each pair correctly adds up to 9.
Inferring/Drawing Conclusions *(weight: 0)*		
Communicating *(weight: 1)*	Student fails to communicate some aspect of the solution.	Student provides any necessary explanation, completes the table, and reports the full results.

VALENTINE HEARTS

TEACHER'S GUIDE

Grade Level: Elementary

Description:

This task assesses student ability to read and absorb a large amount of data and to enumerate possible arrangements of the data consistent with a given set of constraints.

Mathematics:

Math Objects

☐ Number/Quantity ☐ Shape/Space ☐ Function/Pattern

☐ Chance/Data ☒ Arrangement

Math Actions (possible weights: 0 through 4)

| 1 | Modeling/Formulating | 2 | Transforming/Manipulating |

| 2 | Drawing Conclusions | 1 | Communicating |

Assumed Mathematical Background:

This task assumes reading facility at grade level, a basic understanding of arrangements, and some facility with graphic organizers.

Core Elements of Performance:

- Organize a large amount of data.
- Devise a consistent counting scheme that gives a reasonable representation of the described situation.
- Use the model or counting scheme to solve a problem with different constraints.

Using This Task:

Read through the prompt with your students to ensure that they understand the task. Since students are required to read and absorb a large amount of data, it may be helpful for lower-level readers to "act out" the roles of the children and the mothers. At a minimum, the task should be read aloud by either the teacher or students; students should then be asked either to report orally or to write out all the information they remember.

If the task is being used in an informal classroom setting, try to find many different possible ways to organize the data and encourage the students to share their strategies.

Materials Needed:

Colored paper hearts or other colored manipulatives

Name: _____ Date: _____

VALENTINE HEARTS

Mrs. Newman has 5 children: 3 girls named Karen, Lisa, and Amy and 2 boys named Jack and Mark. She is buying chocolate valentine hearts for the children to give to each other. She notices that the hearts are wrapped in either red or blue or green or yellow foil. She decides that she will buy red hearts for the girls to give to the boys, blue hearts for the girls to give to the girls, green hearts for the boys to give to the boys, and yellow hearts for the boys to give to the girls.

1. How many chocolate hearts of each color does Mrs. Newman need to buy? Draw a picture or write a description that explains your answer.

Name: _____ **Date:** _____

2. On the way home, Mrs. Newman meets her neighbor, Mrs. Brown, who also needs to buy valentines for her children to give to each other. Mrs. Brown likes Mrs. Newman's scheme of buying chocolate hearts so much that she decides to use the same rule for doing her shopping. She calculates quickly that she will need to buy 3 red hearts, 6 green hearts, and 3 yellow hearts.

   How many girls and how many boys does Mrs. Brown have? Draw a picture or write a description that explains your answer.

Valentine Hearts   SOLUTION AND RUBRIC

1. Some students may organize the possibilities through a table or a diagram.

   Some students may count without the use of a diagram, and mentally compute 3 girls giving 2 red and 2 blue hearts (a total of 6 red and 6 blue) and 2 boys giving 3 yellow and 1 green heart (a total of 6 yellow and 2 green).

2. It is possible to interpret the statement as meaning that no information is available about the number of blue hearts. Although this temporarily makes the problem more complicated, the result turns out to be exactly the same. Clearly, there will be some number of boys and some number of girls, since red and yellow hearts change hands. For the exchange of red hearts, there are two possibilities: 3 girls give them all to l boy or l girl gives them to 3 boys. If the first possibility is used, there would be no green hearts changing hands. In the latter case, there would be 6 green hearts, which corresponds to the given information. Checking how many hearts would be needed for 1 girl and 3 boys, we find that 3 red, 3 yellow, and 6 green are the appropriate numbers.

	Partial Competency	Full Competency
Modeling/ Formulating *(weight: 1)*	Devise some sort of counting scheme that gives a reasonable representation of the described situation in **1**.	Devise a consistent counting scheme that gives a reasonable representation of the described situation in **1** and **2**.
Transforming/ Manipulating *(weight: 2)*	Implement the counting scheme correctly in **1**.	Work with the scheme in reverse in **2**.
Inferring/Drawing Conclusions *(weight: 2)*	Make some attempt to use the model from **1** to answer **2**.	Successfully used the model invented in **1** to solve the problem in **2**.
Communicating *(weight: 1)*	Give an incomplete or unclear set of answers.	State all answers clearly, whether in prose or diagram form.

MILLIE & MEL'S

TEACHER'S GUIDE

Grade Level: Elementary

Description:

This task assesses student ability to organize information and to enumerate arrangements determined by Cartesian products.

Mathematics:

Math Objects

☐ Number/Quantity   ☐ Shape/Space   ☐ Function/Pattern

☐ Chance/Data   ☒ Arrangement

Math Actions (possible weights: 0 through 4)

[2] Modeling/Formulating   [2] Transforming/Manipulating

[3] Drawing Conclusions   [2] Communicating

Assumed Mathematical Background:

This task assumes a basic understanding of combinations and arrangements, and some facility with graphic organizers.

Core Elements of Performance:

- Come up with a consistent and systematic counting scheme.
- Correctly compute all results, consistent with respect to choices made in counting.
- Present a reasoned, clear argument in defense of your position.

Using This Task:

Read through the prompt with your students to ensure that they understand the task. If students have not worked recently with combinations and arrangements, it will be helpful to do a pre-activity with a simple situation, such as how many outfits can be made from four different pants and three different shirts. Encourage students to share various approaches and strategies as they organize the data; some may use a tree diagram, others a T-table, others a pictogram.

It is important that students eventually come to the realization that the choice of garnish is the key element in the different answers. Many students will come up with a Cartesian product of "bread × garnish × filling" as the most efficient means of solution.

Extension:

Question 4 may be further extended to a larger number of possibilities in each category, or to allowing garnishes or breads to be mixed in the same sandwich. These extensions will force the issue of using multiplication rather than a laborious counting method of solution.

Name: _____ Date: _____

Millie & Mel's

Your Aunt Millie and Uncle Mel have just opened a sandwich shop. Here is a picture of the sign they have hanging behind the counter.

Millie & Mel's Sandwiches		
Bread	**Garnish**	**Filling**
rye	lettuce	salmon
white	tomato	roast beef
French		ham
		tuna
Fresh ingredients	*All sandwiches made to order*	

1. How many different kinds of "tuna on rye" sandwiches can be made? What are they? (You may only order **one** filling and **one** kind of bread in each sandwich.)

2. How many different kinds of ham sandwiches can be made? What are they?

Name: _____   Date: _____

3. Aunt Millie and Uncle Mel want to print a menu listing all the different sandwiches they can make.

Aunt Millie claims they can make **24** different sandwiches.

Uncle Mel claims they can make **36** different sandwiches.

Their best customer claims they can make **48** different sandwiches.

Who do you agree with and why?

Extension:

A new sandwich shop has opened down the street from Millie & Mel's. Here you may choose from 5 kinds of bread, 8 fillings, and 3 garnishes. If you bought a different sandwich in this shop five days a week for your lunch, how many years would it take to order one of each possible sandwich?

MILLIE & MEL'S

SOLUTION AND RUBRIC

All of the questions are subject to interpretation of what choices of garnish are available: (1) either one of the garnishes is available for each sandwich, (2) both garnishes could be combined on a sandwich, or (3) a sandwich can have no garnish.

1. The answer to this question will depend entirely on the choice made about available garnish: either 2 kinds (with lettuce or tomato), 3 kinds (with lettuce, or tomato, or lettuce and tomato), or 4 kinds (lettuce, or tomato, or lettuce and tomato, or no garnish).

2. For each choice of bread there are 2, 3, or 4 choices of garnish. Since there are 3 possible breads, there are 6, 9, or 12 possible combinations.

3. This question gets directly at the idea of how the garnishes are counted.

    According to **Aunt Millie**, there can be 4 (salmon, roast beef, ham, tuna) × 3 (rye, white, French) × 2 (lettuce, tomato) = **24** different sandwiches.

    According to **Uncle Mel,** there can be 4 (salmon, roast beef, ham, tuna) × 3 (rye, white, French) × 3 (lettuce, tomato, nothing added **or** lettuce and tomato) = **36** different sandwiches.

    According to the **customer**, there can be 4 (salmon, roast beef, ham, tuna) × 3 (rye, white, French) × 4 (lettuce, tomato, lettuce and tomato, no garnish) = **48** different sandwiches.

Extension:

Assuming that the 3 garnishes are counted as yielding 7 possibilities, there will be 8 fillings × 5 breads × 7 garnishes, or 280 different sandwiches. It would take a little over a year to eat a different sandwich every weekday.

	Partial Competency	Full Competency
Modeling/ Formulating *(weight: 2)*	Attempt to make an exhaustive list of kinds of sandwiches, with no evidence of systematic organization.	Come up with a consistent and systematic counting scheme, (most probably a Cartesian product of bread × garnish × filling).
Transforming/ Manipulating *(weight: 2)*	Correctly compute some of the results.	Correctly compute all of the results and have all answers consistent with respect to choice of garnish possibilities.
Inferring/Drawing Conclusions *(weight: 3)*	Perform a complete, consistent analysis of either the two or the three possibilities for garnish.	Perform a clear, consistent analysis of the four possibilities for garnish.
Communicating *(weight: 2)*	List results, but present no explanation or defense for any of the answers.	Present a reasoned argument in defense of any one of the positions and clearly communicate that the choice of garnish is the key element in the different answers.

References

Education Development Center. (1975–1981). *TORQUE Project Materials*. Newton, MA: Author.

National Council of Teachers of Mathematics. (1995). *Assessment standards for school mathematics*. Reston, VA: Author.

National Council of Teachers of Mathematics. (2006). *Curriculum focal points*. Reston, VA: Author.

Nuffield Mathematics Project. (1969). *Problems: Green set*. London: Newgate Press.

Schwartz, J. L., & Kenney, J. M. (1995). *Assessing mathematics understanding and skills effectively*. Cambridge, MA: Harvard Graduate School of Education.

Schwartz, J. L., & Kenney, J. M. (1999). *MCAPS: Mathematics content and process scoring*. Cambridge, MA: Harvard Graduate School of Education.

Thyer, D. (1993). *Mathematics enrichment exercises, a teacher's guide*. London: Cassell Press.

Zhitomirsky, V., & Shevrin, L. (1987). *Maths with mummy*. Moscow: Raduga Publishers.

Index

Actions, 3–4
   aspects of, 3–4
   categories of, 3
   mapping to NCTM, 5–8 (figure)
Add 'em up (number and quantity), 52–57
Add-rings (number and quantity), 17–21
Algebra, mapping to NCTM, 6 (figure)
Alphabet coding, 151–154
Arrangement
for elementary grades, 175
   Millie and Mel's task, 188–191
   mixed-up socks task, 164–168
   overview of, 3
   postal puzzle task, 177–180
   for primary grades, 175
   shares task, 181–183
   skills for, 175–176
Assessment. *See* Balanced assessment
Assessment Standards for School Mathematics (NCTM), 1, 12–13

Balanced assessment
   characteristics of, 12–13
   goals of, 1
   mapping to NCTM, 5–8 (figure)
   for number and quantity, 15–16
   rubrics for, 11–12
   tasks for, 10–11
   task weighting, 9–10
Beach day (chance and data), 156–158
Birthday cupcakes (number and quantity), 22–25
Broken calculators (number and quantity), 66–68
Broken measures (number and quantity), 75–78

Calendar moves (pattern and function), 131–142
Chance and data
   beach day task, 156–158
   for elementary grades, 155
   mapping to NCTM, 8 (figure)
   measure me task, 169–173
   mixed-up socks task, 164–168
   overview of, 3
   for primary grades, 155
   skills for, 155
   we scream for ice cream task, 159–163
Coding the alphabet (pattern and function), 151–154
Communicating, 10
Content vs. process, 2
Counting off (number and quantity), 79–82
Curriculum and Evaluation Standards for School Mathematics (NCTM), 1
Curriculum Focal Points (NCTM), 1, 5–8 (figure)

Data analysis/probability, mapping to NCTM, 8 (figure)
Does it fit (shapes and space), 111–113
Dot-to-dot (number and quantity), 33–37

Elementary grade skills
   arrangement, 175
   chance and data, 155
   number and quantity, 16
   pattern and function, 123
   shape and space, 95

Fermi four (number and quantity), 63–65
Function. *See* Pattern and function

Garden's of delight (shapes and space), 119–122
Geometry, mapping to NCTM, 7 (figure)
Grassy parks (shape and space), 97–100

Hallway number line (number and quantity), 29–32

Ice cream task, 159–163
Inferring/Drawing conclusions, 9

Learning expectations, 5–8 (figure)
Leopard's leap (number and quantity), 49–51

Make a map (pattern and function), 143–150
Maps, 143–150
Marigolds, 38–41

Mathematics
  actions in, 3–4
  objects in, 2–3
  structure in, 2, 4
MCAPS: Mathematics Content and Process Scoring (Schwartz & Kenney), 11–12
Measure me (chance and data), 169–173
Measuring marigolds (number and quantity), 38–41
Millie and Mel's (arrangement), 188–191
Mirror, mirror (shapes and space), 114–118
Mixed-up socks (chance and data), 164–168
Modeling/Formulating, 9
Multiplication rings (number and quantity), 58–62

National Council of Teachers of Mathematics (NCTM), 1, 5–8 (figure)
Network news (number and quantity), 83–87
Number and quantity
  add 'em up task, 52–57
  add-rings task, 17–21
  birthday cupcakes task, 22–25
  broken calculators task, 66–68
  broken measures task, 75–78
  counting off task, 79–82
  dot-to-dot task, 33–37
  for elementary grades, 16
  fermi four task, 63–65
  hallway number line task, 29–32
  leopard's leap, 49–51
  mapping to NCTM, 5–6 (figure)
  measuring marigolds task, 38–41
  multiplication rings task, 58–62
  network news task, 83–87
  overview of, 2
  piece of string task, 88–93
  postal puzzle task, 177–180
  for primary grades, 15
  shares task, 181–183
  table talk task, 42–48
  trouble with tables task, 69–74
  TV show schedules task, 26–28

Objects, 2–3
  arrangement, 3, 175–176
  chance and data, 3, 155
  mapping to NCTM, 5–8 (figure)
  number and quantity, 2, 15–16
  pattern and function, 3, 123–124
  shape and space, 2, 95–96
  and tasks, 9–10

Pattern and function
  calendar moves task, 131–142
  coding the alphabet task, 151–154
  for elementary grades, 123
  make a map task, 143–150
  mapping to NCTM, 7 (figure)
  overview of, 3
  for primary grades, 123
  skills for, 123–124
  wall design task, 125–130
Piece of string (number and quantity), 88–93
Postal puzzle (arrangement), 177–180
Primary grade skills
  arrangement, 175
  chance and data, 155
  number and quantity, 15
  pattern and function, 123
  shape and space, 95
Process dimensions, 2
Process vs. content, 2

Quantity. *See* Number and quantity

Rubrics, 11–12
  See also specific task rubrics

Shape and space
  does it fit task, 111–113
  for elementary grades, 95
  grassy parks task, 97–100
  mapping to NCTM, 7 (figure)
  mirror, mirror task, 114–118
  overview of, 2
  for primary grades, 95
  shirts in the mirror task, 104–110
  skills for, 95–96
  stickers task, 101–103
Shares (arrangement), 181–183
Shirts in the mirror (shapes and space), 104–110
Skills
  arrangement, 175–176
  chance and data, 155
  number and quantity, 15–16
  overview of, 9–10
  pattern and function, 123–124
  shape and space, 95–96
Stickers (shape and space), 101–103
Structure, of mathematics, 2

Table talk (number and quantity), 42–48
Task weighting, 9–10
Teacher's guides, overview of, 11
Transforming/Manipulating, 9
Trouble with tables (number and quantity), 69–74
TV show schedules (number and quantity), 26–28

Valentine hearts (arrangement), 184–187

Wall design (pattern and function), 125–127
Weighting, of tasks, 9–10
We scream for ice cream (chance and data), 159–163

The Corwin Press logo—a raven striding across an open book—represents the union of courage and learning. Corwin Press is committed to improving education for all learners by publishing books and other professional development resources for those serving the field of PreK–12 education. By providing practical, hands-on materials, Corwin Press continues to carry out the promise of its motto: **"Helping Educators Do Their Work Better."**

In compliance with GPSR, should you have any concerns about the safety of this product, please advise: International Associates Auditing & Certification Limited The Black Church, St Mary's Place, Dublin 7, D07 P4AX Ireland
EUAR@ie.ia-net.com

www.ingramcontent.com/pod-product-compliance
Lightning Source LLC
Chambersburg PA
CBHW081355290426
44110CB00018B/2383